AF783093

Transforming IT

A Five-Step Plan for Adopting a Service-Centric IT Model

By Keith Jahn and Rebecca Lawson

Copyright © 2008 Keith Jahn and Rebecca Lawson

www.transforming-it.com

All Rights Reserved

No part of this book may be reproduced in any form or by any electronic or mechanical means including information storage and retrieval systems, without permission in writing from the authors.

ISBN 978-1-4276-3216-6

Third printing August 2008

Acknowledgements

As an IT practitioner and strategist, I have been blessed to be surrounded by many highly energized and brilliant people. I am thankful for willing and capable mentors who trained and shaped me professionally: Dave Dutton, Jeff Groner, Cloyd Wempe, Ken Shumate, Mike Baker, Russ Daniels, and Nora Denzel.

I would like to acknowledge and thank my professional colleagues who allowed me to witness and participate in achieving extraordinary feats in the face of unbelievable circumstances which serve as a backdrop for this content: David Bills, Vercie Lark, Ray Webb, Bruce Meier, Mike Baker, Mary Carol Kelley, Renee Dion, Larry Ketchersid, Glenn Osterhaus, and Larry Rushing.

I would like to acknowledge some amazing CIOs, IT and outsourcing leaders who afforded me opportunities that exposed the various aspects of how IT works in large-scale corporations: John W. White, Michael Capellas, Bob Napier, Vallerie Parrish-Porter, Tami Mallett, Randy Burdick, Pete Karolczak, Randy Mott, and Uli Holdenried.

Two people had a profound impact on me personally and professionally and to whom I dedicate this book. Words do not do justice to express the extent of my gratitude. First, Dave Scroggins, who for almost every day of 15 years stood by me, challenged my ideas and picked me up whenever I stumbled—which was quite often. No one will ever fully understand the depth and breadth of the impact Dave had on my life and my career. I miss my friend. And second, Fred Jones, who had more wisdom, insight, energy and tenacity than anyone I ever worked for. He had more faith in me than I had in myself, nurtured me like a father, and pushed me harder than any coach I ever had. This book would not have happened without his constantly assigning me new and challenging opportunities.

And finally, to my extraordinary wife for believing in me, supporting me, and letting me spread my wings into this endeavor with total support and confidence.

-- Keith Jahn

For the light and love and laughter that energizes me day after day, I'd like to thank Katherine, Roman, Paula, Carla, Chris, Liz, John, Matt, Robin, Zoe, and Georgie.

A special thanks to Emile Barrios for helping with the structure and copywriting of this book.

-- Rebecca Lawson

Contents

Introduction

Just Another Day at the Office…

It's 1:31 a.m. Tuesday morning in the monitoring wing of ACME corporation's data center. The third shift operations team is on site. Processing of batch schedules is in progress and, as usual, it's going to be a race against the clock. More than 7,200 batch processing routines must execute across 250 systems, pushing and pulling and massaging almost six terabytes of information. The entire process must be completed within a twelve-hour window so that ACME's employees have what they need to do their jobs on Wednesday.

The operations team is not looking happy. For the past two hours the batch schedules for manufacturing have been waiting on a single parts inventory file to arrive from a major supplier in Asia. This file is critical for compiling the factory build schedules to be used by the assembly lines in the morning. Processing the file typically takes about an hour. If the batch job that produces the factory schedules isn't complete by 3:00 a.m., manufacturing will be delayed by eight hours and the entire first shift of 280 workers will be sent home. If that happens, customer orders will be delayed by at least a day, which will cause order cancellations, missed commitments to transportation carriers and lost revenue—not to mention repercussions from unhappy and unpaid employees.

The batch scheduling staff has been trying to get an estimated time of arrival from their IT contacts in Asia, but have not yet received any response. Standard operating procedures call for the manager on duty to escalate the problem to the VP of operations once the processing window threshold of 1:30 a.m. has been exceeded. The call is made.

Sleep-deprived and weary from the prior nights' disruption associated with a corrupted database in the ERP environment, the VP is up once again and on the phone attempting to alert and solicit help from the director of inventory management, who has just drifted off to sleep after spending an evening in the emergency room with his six-year-old son who took a nasty spill after dinner while learning to rollerblade. After being apprised of the situation, the director of inventory management proceeds immediately through his escalation channels, given the time of day and the magnitude of the situation. Using the speed dial on his Blackberry, he is able to get through to the Asian supplier. The file finally arrives at 2:00 a.m.. The batch schedules begin executing again by 2:03, but the team knows the delay will keep tensions high until morning.

While this drama is unfolding, the server administration team launches a pre-approved set of changes to implement a set of security upgrades for the 450 servers in the manufacturing environment. They begin at their scheduled time, 1:30 a.m. IT personnel from the system administration team are on-site to ensure expediency and success of the upgrades, given the complexity of the environment and sensitivity to timing…it's the last week of the fiscal quarter. These upgrades require a reboot of each system which generates a 12-15 minute downtime window for each application running on the server.

Just as he is drifting off to sleep again, confident that the manufacturing crisis has been averted, the VP of operations sits straight up in bed. Heart racing, he reaches for the phone and calls the lead system administrator who is on site at the data center. She reports that half of the 450 servers have received the upgrade, but the work is on hold because of problems with several servers that haven't rebooted properly. The issue appears to be network-related, she says, but the monitoring systems aren't showing any network outages in the environment. The network engineer on call has been engaged to troubleshoot the problem.

The VP of operations explains the situation with the batch schedules and instructs the system administrator to delay the remainder of the upgrades until after inventory processing has completed. The system administrator reminds him that this time window was the only one they could negotiate with the business and if they don't get it done by 3:30 a.m. they won't be able to get another window until the end of the next quarter, leaving the systems exposed to known security vulnerabilities that have been exploited and caused business downtime in other environments.

Knowing that the systems that have not rebooted properly are instrumental in the factory build schedule process, the VP conferences in the batch schedulers who are indeed not seeing the response they would expect from the processing systems. The network technician joins the conference call and explains that there does appear to be a problem with the router in the manufacturing environment which will require a reboot - taking all 450 systems offline and further delaying batch processing for at least ten minutes.

As the clock nears 2:30 a.m. the reality of the situation settles in on everyone involved: the build schedules aren't going to be delivered on time.

Taking a deep breath, the VP of operations decides to make the call. He wakes the CIO and brings him up to speed on the situation. The director of inventory decides to alert the supply chain VP, along with the factory manager. The supply chain VP is livid that this is happening at the end of the quarter when shipments are critical to meet the company numbers. He can't understand why IT made the decision to perform such disruptive tasks during this critical period.

The discussion rapidly deteriorates into an attack on IT's competency and lack of understanding of how the business operates. The conversation ends when the supply chain VP says to the CIO: "You can be sure of one thing - when we miss our numbers for the quarter and make the front page of the *Wall Street Journal*, you and your IT systems failures will carry the blame."

The Elusive Goal: Running IT Like a Business

While there are several company-specific problems in this scenario ranging from outdated system design to lack of change governance, if you're an IT executive, manager or professional working in a large corporation, a situation like this is probably very familiar to you.

Why has this scenario become the normal (and seemingly permanent) state of affairs in IT departments worldwide?

For at least a decade, IT leaders and technology suppliers have focused on reshaping IT organizations to be more nimble and more capable of delivering valuable business outcomes in a predictable, cost-effective manner. So after all the time, effort and money spent on improving the way IT operates, why does this goal still seem to be a utopian dream?

One reason is the central role IT now plays in the modern enterprise: a universal and indispensable mechanism for enabling every facet of the business. IT helps organizations achieve greater customer accessibility and responsiveness. IT enables virtually every employee in every department do his or her job better, and it gives the enterprise the visibility it needs to properly monitor and control its component parts across vast distances. IT is the obvious first choice for handling these kinds of business needs because implementing technology is usually the most powerful and most cost-effective way of doing so.

But as any CIO will tell you, managing technology and managing the IT organization are two very different things. Technology by its nature gets more complex over time. Additional features and capabilities enable the business to use technology to solve larger and more difficult problems—in other words, to become more dependent on IT for day-to-day operations. But with IT reaching now into the smallest and most remote corners of the enterprise, the task of managing and operating those technologies in the ever-changing context of the businesses they support has gone from difficult to almost impossible.

Think about it: the largest percentage of the average corporate IT budget is spent simply to maintain and operate core business systems—leaving very little room for investment in business innovations.

Among a growing stack of unfunded initiatives are the organizational improvements IT itself needs to keep up with the pace of business change. Yet CIOs, like the one at ACME Corporation, must constantly deal with the perception that IT is nothing more than a cost center and an inhibitor to the business, instead of a value-based enabler helping the organization achieve its financial and performance objectives.

So while the business moves inexorably toward greater and greater reliance on technology, it continues to treat the IT organization as an out-of-control teenager whose allowance must be reined in at every opportunity. This paradox has caused a major shift in the debate about the place IT occupies, from a simple discussion of how to provide enabling technology to how the business can leverage IT for maximum value and competitive advantage in the marketplace.

The debate is generally triggered by conversational questions such as:

- What, exactly, does IT do for my business? I can't see IT driving any business innovation - I just see cost. What do I get for the huge IT allocation I see every quarter?

- Why does it take so long to simplify and consolidate IT operations? Why are the investments required so high? IT costs need to come down and the quality of deliverables needs to come up by next quarter!

- Why should we fund IT projects when most of the efforts we do fund aren't delivered on time, on budget, or even resembling the original request?

- We're trying to grow the business, but IT is in the way. Why can't they keep up with the pace?

Questions like these reinforce the fact that the business still views IT as a supplier instead of enabler of business value. This increases the pressure on IT to achieve new levels of maturity, capability, performance and cost structure transparency.

But the structure of IT as it exists today is, frankly, inadequate to this task. IT spends so much time and money trying to meet existing business demands that it does not have the resources to focus on improving its internal operations or aligning itself with the business it serves.

> IT must reinvent itself as a sustainable, service-oriented line of business focused on delivering business value.

So in order to address the issue, IT must reinvent itself as a sustainable, service-oriented line of business focused on delivering business value. This requires making significant changes to IT's structure—changes which traditionally are the most difficult to make and the easiest to avoid. Making these structural changes also requires making changes in corporate attitudes about IT, as well as the allocation of time and money.

In order for this wholesale re-positioning to work, IT must move out of the computer room and into the boardroom. This isn't just the CIO's strategy, it's the strategy for the entire business.

The capability-centric model

As any business matures, the operating models it uses are recalibrated to meet the changing needs of the business. That's a very sensible strategy, but it is a strategy rarely applied to the IT organizations within those businesses. Instead, IT organizations generally evolve around the delivery of technology-based capabilities (like servers, networks and storage) in response to the company's current needs and business processes. As new IT capabilities come online, they simply are added onto the list of "Things That IT Provides".

Under this capability-centric structure, IT struggles to achieve a higher level of maturity because it is too busy handling the everyday job of providing and managing the universe of technology and support resources that business operations require. Making matters worse, this capability-centric approach usually results in a fragmented organizational structure, held together by people rather than by deliberate organizational design. The people of IT divide themselves into "tribes" or siloed groups, each with its own culture, rituals, language and tools for producing and managing their specific capabilities.

A key weakness of capability-centric IT is the way it addresses business needs in a build-to-order (or engineer-to-order) manner, rather than a configure-to-order approach. In fact, capability-centric IT actually *causes* this build-to-order mentality.

Using the build-to-order approach, individual projects or requests are fulfilled by constructing unique solutions from non-standard elements. This demands human labor, longer lead times and a higher margin of error. The technology of these one-off projects usually can't be leveraged for other uses, and they are expensive to develop and maintain over time. In contrast, the configure-to-order approach fulfills requests by configuring standardized solutions from a finite set of predefined elements.

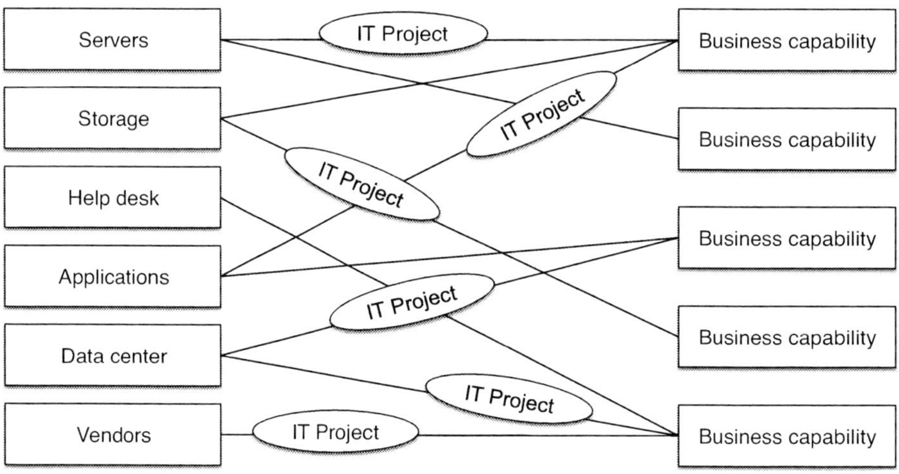

Figure 1 - Capability-centric IT results when IT is organized around siloes. At best it can deliver IT projects that are tactical, expense, and typically build-to-order by design.

In most IT organizations, the vast majority of solutions are build-to-order, which in fact is a major reason why costs keep spiraling out of control and IT's reputation as a budgetary black hole is perpetuated.

At best, capability-centric IT organizations can use a transition strategy to create some improvements, but these organizations will continually struggle to achieve a transformed end state where the unit of measure and basic organizing principle is centered on the actual services delivered to the business. The reality is that most transition initiatives in a capability-centric IT organization will fail to produce the desired end state. You can't re-invent an IT organization by updating and upgrading pieces here and there while leaving the underlying behaviors, techniques and operating models intact.

The service-centric model

Service-centric IT represents a state of maturity for the IT function, enabling it to operate as a service-focused entity which is deliberately structured, organized and calibrated to directly power the corporation's strategic growth and profitability objectives.

A service-centric IT organization is fully aligned with the business it serves, because its structure and operating model are defined in the context of that business. The technology-enabled services are designed, packaged and offered to the lines of business to meet the specific needs of users, and have an associated value and cost that both sides can easily understand. Users of each service are willing to pay that cost to receive that value, and everyone from the cubicle to the executive suite can see it and measure it.

Adopting a structure and operating model for IT centered on the service as the complete and irreducible unit of value produced by IT addresses both the "what" of IT (its services) and the "how" of IT (the design, construction, delivery, cost and value of these services).

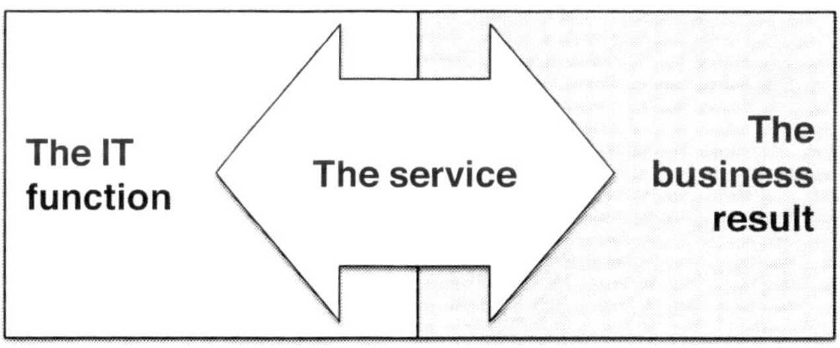

Figure 2 - The service is what connects IT with the business

In the service-centric IT organization, there is no question of what the business gets for the money spent on IT, because IT has achieved its ultimate objective—to be accepted as a true value-based driver of business results.

Transition and Transformation

To achieve an end state where IT occupies its proper strategic role, IT must leave behind the notion of *transitioning* to new or improved process and technologies and adopt a completely different kind of strategy—a *transformation* strategy. Such a strategy changes the fundamental structure of

IT to reflect how the business actually operates, and to allow IT to assume its rightful place as a valued partner to the enterprise.

Before we go any further, let's define some terms that set context for this discussion.

Transition is the progression of the current state towards an end state vision in the context of existing structures, methods and behaviors. Transition is evolutionary in nature, where a set of projects or initiatives are continually revisited based on available funds and/or priorities of the day. Transitional efforts are generally driven by technology catalysts or by best practices. Because of their weight, most strategic efforts atrophy when approached in a transitional manner, and while some goals are achieved, they typically are at a surface level—the strategic end state never is attained. Importantly, a transitional approach to change does not address the fundamentals of the IT function, which include its structure, behaviors, and outcomes.

Transformation is a method of progression toward a strategic end state vision which encompasses new techniques, behaviors and methods for executing IT's activities. The strategic end state, which is significantly different from the current state, delivers lasting change. Transformation produces technology and process improvements, consolidation and simplification.

While the idea of transforming IT is not new, it is clear that few companies are actually doing it successfully. Not surprisingly, the toughest parts of transformation are clearly articulating what the end state should look like, and changing the way things are done—because people by nature resist change. But you can be sure that it is indeed possible and vital to undergo this transformation in ways that are both incremental and easily achievable.

The purpose of this book is to describe a model for how this can be done.

Very few capability-centric IT shops are making measurable progress toward transforming into a service-centric shop. More often, these organizations are focusing on transitional tactics aimed at the endless pursuit of cost reduction: off-shoring, automation and consolidation, among others. The result is a kind of "infinite loop" state, where the same issues are addressed over and over.

This is why the list of top ten CIO concerns published by industry analysts doesn't change much from year to year. This list includes things like business and IT alignment, cost reduction, leadership, infrastructure refresh, and security. The items might be re-arranged each year, but the list is still the same, because transition initiatives in capability-centric IT organizations simply don't attack the root cause of IT's problems. Instead of dealing with these root causes, more technology is thrown at the problems—which, of course, makes things worse.

For example, consider the virtualization of computing platforms. Virtualization is itself a disruptive technology in that it requires IT to organize and work in new ways. However, in a capability-centric IT shop, virtualization creates a huge new problem: how to deal with both the legacy and the current state environments along with the new virtualized environments. Given a finite (and, in many cases, shrinking) level of resources, this new virtualized environment doesn't create a transformed state, it merely introduces a new method of computing—and IT is left with the job of managing and supporting multiple architectures simultaneously, because not everything can or will be virtualized. In addition, the core processes relating to these environments must be augmented, updated and supported, instead of being replaced.

If all of this is sounding familiar to you, don't despair—you're in good company. It's the same battle IT executives have been fighting for years. Now, finally, there is a solution.

The solution is to take the IT function and transform it into a new state, where focus is the deliverables IT provides to the company's lines of business rather than the capabilities it is required to support. These deliverables are *IT-provided services*, the product line of IT. Using the IT service rather than IT capabilities as the central organizing principle enables IT to transform itself around them—which in turn leads to the implementation of next-generation technology platforms with consolidated and automated processes.

We call this approach service-centric transformation.

Creating a New Mindset

The service-centric approach begins with changing the way IT thinks about itself, based on a clear end state that's different that what IT looks like today.

So what exactly does it look like?

No longer a supplier of capabilities, service-centric IT is a service-oriented line of business, where services themselves are IT's deliverables. The customers of IT are the economic buyers or consumers of these services in the various lines of business, who derive some level of value from them and are willing to pay money to obtain and use them. Under this structure, IT products, projects and assets are managed as a set of portfolios, much like stock portfolios.

With this mindset in place, IT can implement a structure and an operating model that focuses on managing supply and demand for its line of services—in other words, concentrating on service management instead of server management. The activities involved in designing, delivering and managing these services are the functions that will, in turn, define the form of IT.

There is nothing new in this; this is the way things work in virtually any other line of business. The product and the customer are the center of activity and focus, and the business model is constructed in a way that addresses customer needs with goods and services in the most efficient, effective manner.

Unfortunately, this useful, proven methodology has seldom been applied to IT in a way that's manageable, sustainable, and replicable.

Creating such an operating model for IT requires value-based design, where project portfolio investments (such as upgrades, expansion or automation) are connected with and measured by business value, and are vetted and approved by the internal communities these investments will serve. Importantly, improvements to IT's ability to produce business value are placed in the context of the overall business, instead of the context of technology alone. The services IT offers its customers move through a series of value chains that shape their quality, consistency and effectiveness.

In undertaking service-centric transformation, IT can use mainstream business lines such as manufacturing as a guide to designing and building a new operating model. In fact, leveraging manufacturing concepts like supply chain, mass customization, and the just-in-time inventory model provide a solid foundation for the end state of IT.

The new service-centric IT model has two objectives:

- **Minimizing business risk** by improving organizational performance and more effective management of the IT project portfolio, along with reducing downtime and exposure to security threats.
- **Maximizing business value** by enhancing IT's ability to deliver services and by reaching a new level of business innovation.

It's important to note that the success of this transformation strategy is largely dependent on the way IT is positioned within the enterprise. Just like

IT itself, the executive committee and the board must also adopt a different mindset—specifically, leaving behind the cost-centric mentality and concentrating on IT's potential to add value and drive the business forward. This means giving a seat at the executive committee or boardroom table to the CIO.

Transforming IT to a service-centric line of business addresses the underlying cause of ineffective IT, and thus delivers lasting benefits and a higher degree of maturity, yielding important characteristics for the organization:

Agility: making IT more flexible so that it can reduce time-to-market for business initiatives and facilitate change

Financial transparency: improving the cost structure of IT to better extract business value from each project and/or service consumed

Control: managing IT more effectively to reduce risks associated with each new project or service

Alignment: linking the structure and operating model IT with those of the business as a whole to improve operational efficiency and relationships between IT and lines of business.

The Journey of Service-Centric Transformation

This book is about how to take (and survive) the journey of service-centric transformation.

Make no mistake, this is a difficult journey, but it is a journey IT *must* take if it wants to successfully provide business value and drive innovation. Existing capability-centric operating models are incapable of scaling to meet the needs of the large enterprise, as the current state of IT clearly demonstrates. Without a transformation strategy, that situation will continue to worsen.

As any IT professional can attest, the specter of failure always looms over initiatives this big and complex. The good news is service-centric transformation can be done in a simple, incremental way that delivers significant and continuous short-term value. This, combined with IT's ability to show substantial ongoing progress toward the desired end state, will serve to quiet criticism and provide growing momentum.

To sum up: growing companies have rapidly-changing IT needs which have rendered the capability-centric IT model obsolete. Using a transition strategy to stay ahead of this curve is like rearranging the deck chairs on the *Titanic*—any short-term benefits are relatively meaningless because over the longer term IT as a whole continues to sink.

Adding maximum value and delivering optimal business outcomes requires that IT make major strategic shifts in its structure and operating model to improve how it is aligned with the business as a whole. This requires positioning IT as a service-centric line of business within the enterprise, with a structure and operating model that is both powerful and complete enough to support the entire business—which, after all, is what the business needs from IT. By re-inventing itself using experiences from mainstream business lines, the IT organization will significantly improve how and how well it powers the business.

Setting Context for Re-positioning IT

Service-centric transformation is a complex, large-scale, time-intensive initiative that involves upending the status quo and changing the way IT operates at the most basic level.

How can that possibly be a good idea?

The answer can be found by looking closely at the way almost every corporate IT department currently operates, and how well that structure and operating model serves the enterprise as a whole. Armed with that knowledge, a new way forward can be found and followed.

Service-centric transformation is that new way forward, which ushers in a new structure and operating model for IT. This structure links IT and the business at a single point of economic control: *the service.*

The service, in this context, is defined as a single, irreducible unit of value delivered by IT and consumed by the business. In other words, an IT service is exactly the same thing as a product on a store shelf: a self-contained package designed to meet the need of a specific market and priced for sale in the context of the value it brings to consumers in that market. Like any product, an IT service can contain a wide range of ingredients: hardware, software, support, technical expertise, etc. Examples of IT services include mobile messaging, payroll, QA lab shared service, supplier access and call center telephony.

This single point of economic control is designed, managed, and measured through a structured enterprise architecture, which includes four value chains (IT administrative management chain, business relationship management chain, design chain and delivery chain,) as well as a structured operating model.

In this section, we begin by evaluating the existing capability-centric model to better understand why it keeps IT from moving into a more strategic role.

Then, we'll explore an alternative four-pillar framing that underpins the service-centric approach.

Capability-Centric IT is Unsustainable

As we previously noted, the list of CIO priorities—including IT business alignment, cost reduction, security, consolidation, outsourcing and staff retention—hasn't changed much over the last ten years. What has changed is that the problems have become so bad as to threaten the very foundation of IT, not to mention the jobs of the CIOs themselves.

IT is oversubscribed, understaffed, and in search for a more sustainable model.

This list of issues persists because these items are *symptoms* of IT's problem, not the problem itself. The root cause of IT's current state is the capability-centric paradigm that can neither scale or flex to meet the exploding needs of today's large enterprise.

Ask most employees in an enterprise how they feel about their IT department and chances are they'll tell you that IT is hard to engage and slow to move. Ask any IT professional in that same enterprise about what their life is like, and he or she will probably tell you they are under-funded, overworked, doing their utmost to keep business systems executing—and feeling like they're failing because the employees do nothing but complain.

What's more, because leaders in the boardroom don't understand the value received from IT, they continue to reject investments that would might improve IT's capability to deliver. Instead, they exhort IT to "do more with less." Alas, the cobbler's children have no shoes.

The pitfalls of a self-fulfilling prophecy

If you look at most enterprise IT shops, you'll find a capability-centric model, focused on delivering X amount of computing, storage and network capability along with basic support to end users in all areas of the business at a cost based on Y percentage of revenue. The cost for these capabilities is generally passed on to the bottom line in some form of allocation or bill. In this model, by the time the bill arrives (if it ever does), the business has lost sight of any real or perceived value derived from the capability. Without any link between cost and value, it's easy for the business to look at the bill as just another out-of-control IT spending spree—and resolve once again to rein in IT's budget.

As time passes, IT needs to invest in new technology and systems to maintain and improve existing capabilities so the business can meet new or changing market demands. But of course, the budget isn't there. So IT is expected to deliver the new capabilities inside the existing cost envelope, which turns the natural state of tension into a full blown conflict.

IT has been evolving in this way for over twenty years, which is why CIOs find themselves under so much pressure to "run IT like a business," and yet struggle to do so.

What does a capability-centric IT shop look like? Generally, it is organized around silos—call them "tribes"—who have little contact with and little concern for the other siloed tribes, and don't have the time or find it necessary to understand the business they are supposed to be serving. The tribe's job is to deliver a capability to someone requesting it. They are generally grouped by specific areas of responsibility or subject expertise: systems and network engineers, database administrators, IT help desk, monitoring centers, application development, and deep technical support teams, among others.

In each group, a specific culture, language and behavior evolves over time ("this is the just way we do things here") as each tribe attempts to resolve its own issues without considering the larger context of IT, much less the business user. Because the tools and data used by each tribe tend to be task-specific, they are not usually shared. There is even fracturing within each tribe. For example, level two support specialists create rules and processes which aren't shared with the less-respected level one techs.

Like any bureaucracy, this collection of siloed tribes drifts away from its original mission over time and focuses more on the enforcement of its entrenched rules.

As a collection of autonomous tribes, the capability-centric IT organization is incapable of adapting to the pace of change inherent in large-scale enterprises. Capability-centric models don't scale well because they are rooted in technology and relationships between people rather than in a deliberately structured, value chain based model. As a result, IT continually finds itself in a defensive position, responding to various identified needs or crises instead of being a proactive enabler of the business as a whole. The fact that this makes it harder for the business to engage with IT (and makes it much harder for IT to deliver what the business needs) becomes a secondary consideration.

Like in the ACME scenario, IT becomes all about the crisis du jour.

IT spending patterns

One way to measure the weakness of capability-centric IT organizations is to look at how IT invests its money.

Over the past two decades, the implementation of technology solutions has revolutionized the way people work, improving speed, agility and productivity by several orders of magnitude. As a result, organizations are today almost completely dependent on technology to keep their day-to-day operations, customer interactions and internal systems running smoothly. As technology evolves, new and innovative solutions become available that can help the enterprise do its even faster and more effectively. Given the organization's dependence on technology, it would seem that one of IT's biggest jobs would be to foster this kind of innovation on a continual basis.

Yet just the opposite is true. Most large entities (after all the arguing about what they get for the money) spend between two and eight percent of revenue on IT. Of this, IT organizations spend about 80 percent simply on operating and maintaining existing capabilities, leaving 20 percent or less for evaluating and implementing the new innovations that the enterprise must have in order to maintain its competitive edge moving forward. This concentration on operations and maintenance spending is a direct function of the capability-centric model.

This dynamic is not lost on executive management, whose predictable response is to force IT into constant cost reduction since no connection can be made between the money being spent by IT and the value being delivered. To these executives, whose careers revolve around delivering value to shareholders, IT projects are not aligned with overall business objectives and have unclear value propositions. IT's silo structure makes horizontal prioritization impossible, so projects are generally managed as wish lists instead of portfolios.

In such a muddled environment, the easiest course of action is to direct IT to reduce its run rate spending. To accomplish this, IT responds by taking a number of well-known approaches, including labor arbitrage (off-shoring), automation, and server and data center consolidation. While this provides temporary executive-suite satisfaction in terms of overall spending, it only serves to exacerbate the problems inherent in the capability-centric model, and engender the same cost-cutting discussion during the next budget cycle.

Build-to-order services

Another limiting dynamic in capability-centric IT organizations is the build-to-order vs. configure-to-order approach.

As previously noted, build-to-order means that individual requests are fulfilled by constructing unique solutions from non-standard elements. This happens all the time in IT, especially in projects funded by a specific business unit. Configure-to-order, on the other hand, means that requests are fulfilled by configuring solutions from a pre-defined set of standardized elements.

One reason why eight out of ten IT solutions are build-to-order is the allure of flexibility and the misconception that such an approach is actually affordable. While these one-off projects might be well-funded on the front end from a business unit, and can be seen as delivering specific and measurable value for the money, they often have down-stream financial and support ramifications that are not fully considered. This is especially true in terms of a particular solution's potential for reuse in other areas.

In reality, the business doesn't understand what it is paying for, IT has a hard time getting straight answers from the tribes about where all the money is going, cost allocation models don't work, and the perceived satisfaction level of IT as a whole sinks lower and lower.

Hub-and-spoke management model

Capability-centric IT organizations often are managed as a collection of cost-focused, technology-oriented groups. This creates a hub-and-spoke model in which the spokes (the tribes of IT) are vertically oriented and poorly integrated.

The hub of this model is generally embodied in a workflow engine, most commonly the IT ticketing system.

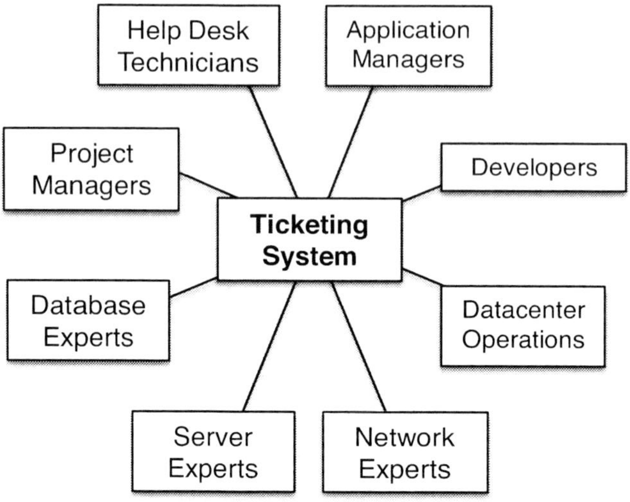

Figure 3 - Hub-and-spoke model with the ticketing system environment as the integration point

The ticketing system functions as a kind of Rosetta Stone for inter-tribal communication and connecting disparate processes. As the business gets larger and demands on IT increase, the ticketing system becomes the critical component for tracking events and issues across the enterprise. Over time, the ticketing system is modified to extend into various parts of the business, and in effect becomes the front door for IT engagement—which is not a good thing.

Independent tribal processes (the bureaucracy each tribe invents to ensure successful execution of its daily duties) are codified in the ticketing system. It doesn't take long for this process to create a highly-complex IT environment consisting of disparate capabilities, rituals, languages and tools. Costs are high, and there is a great potential for error due to the overwhelming dependency on human labor to fulfill the build-to-order solutions that users request through the ticketing system hub. Meaningful automation of horizontal workflow is virtually impossible.

Under these circumstances, it's easy to see that by operating under the capability-centric model, IT organizations will continue to lose ground in their fight to provide the outcomes needed to achieve overall business objectives.

Why Transitional Initiatives Fall Short

CIOs and their management teams spend untold amounts of time, effort and money trying initiative after initiative to better adapt to the changing business landscape. But these initiatives keep failing because they reflect a transitional approach rather than a transformation approach to the strategy—making adjustments to technology, facilities, staffing etc. inside the existing structure rather than reconstructing them around a deliberate, well-designed end state model.

Transition initiatives generally fall into three categories:

- *Cost reduction* through labor arbitrage, consolidation or automation
- *Improved business alignment* through greater process consistency and portfolio management
- *Changes in governance*, such as enterprise architecture, asset sharing to improve utilization, and virtualization.

Right now, most large companies are executing or planning to execute activities in each of these areas. While delivering some gains, the IT professionals in these companies already know that the end result will fall short of expectations. The four most common reasons why transitional IT initiatives fail are: lack of time, investment constraints, lack of planning, and changing business priorities. Let's look at each one.

Lack of Time: Strategic IT programs are by nature very large and can take three years or longer to fully complete. Continuous execution of such an initiative over that period requires a dedication and rigor for which most IT shops don't have the stomach. Time becomes the initiative's worst enemy. Strategic efforts are made up of multiple projects, some short-running like application development, others longer-running like data center consolidation. With so many moving parts, it's easy for IT to lose focus and momentum as time goes on. Finally, there is the fact that initiatives are often abandoned after supporting leadership departs; in other words, initiative time horizons are longer than your average CIO tenure.

Investment constraints: Strategic efforts require investment discipline, which must stand the test of the time horizon mentioned above, and include both capital and operating expense components. Because IT is generally seen as a cost center that generates overhead to the bottom line, the enterprise expects immediate gratification from the initiatives. Too often, though, companies invest for the short term in these projects and pirate any savings for bottom-line reporting purposes. Ideally, these savings should be re-invested for future phases of the project in order to minimize the impact on corporate P&L targets.

Lack of a plan: "Service-centric transformation" and "running IT like a business" can both sound like very opaque objectives. Thus, senior management often doesn't fully understand the relationship between the initiatives that IT is undertaking and how they related to the overall objectives of the business. This problem is compounded in many cases when IT executives rely on their staff to fill in the details required for execution. Since the staff is already overworked, there is no time to formulate effective plans to achieve stated objectives—with resulting confusion and hostility in the boardroom. In order to succeed, IT must first understand what IT transformation really means, and then be able to apply that definition to create strategies and tactics that can be spelled out very clearly in terms that both IT and senior management can understand.

Changing business priorities: Almost every business event triggers an IT action. But too often, changing business needs result in changing priorities within IT. For example, changing economic climates may call for shifting resources to other focus areas. In such a case, IT priorities must be viewed on a par with those of other business functions within the corporation. If the company's boardroom players don't understand and agree with that idea, IT will quickly make its way to the top of the list for budget cutting, and its mission will fall well short of expectations.

So, IT's attempts to strategically reinvent itself through transitional activities reach a point of atrophy, or do not meet their stated expectations upon completion. This happens because the transition initiatives are not focused on changing behaviors and are so ambiguous that the company loses interest or runs out of people or money before they can be completed.

Because true transformation initiatives require shifts in both IT and the business, CIOs and their business partners often de-prioritize and drop funding for them, only to start over again one or two years later.

Service-Centric IT as a Sustainable Model

Given the inherent weaknesses and the historical underperformance of the capability-centric model, IT leaders must now accept the fact that this model is never going to take the organization where it needs to be—no matter how much money is spent or how many transition initiatives are tried or completed.

Again, a service-centric IT organization recognizes that the "service" is the deliverable to the line of business. Each service is defined, designed, delivered and managed through four distinct value chains that comprise and define the IT organization mechanics. These value chains provide a consistent and economically balanced pipeline of services that support business execution and drive business innovation through technology and information.

Underpinning the value chains is a well-defined operating model, stratified into process models, information models, automation and resource layers. The operating model and value chains interlock to govern and orchestrate the complete lifecycle of each service delivered to the business.

Simply said, service-centric IT applies a disciplined approach to abstracting the various IT silos into a single, cohesive "IT factory" centered around business metrics, end user value orientation, and a balanced supply chain mentality to deliver business outcomes.

Simply said, service-centric IT applies a disciplined approach to abstracting the various IT silos into a single, cohesive "IT factory" centered around business metrics, end user value orientation, and a balanced supply chain mentality to deliver business outcomes.

Outcomes of the service-centric IT model include:

- A disciplined and productive engagement model that allows IT and the lines of business to focus and agree on business outcomes
- IT is empowered and structured to deliver services in a predictable, high quality, cost-effective manner
- Lines of business develop a greater trust in IT and what the function does for them. They are more inclined to engage IT for the design, delivery and support of services that drive innovation and competitive advantage
- Supply and demand can be rationalized and calibrated based on quality, risk and measurable value to the business
- Cost-to-value ratios for IT services are both transparent and measurable

The Transformational Path to Service-Centric IT

Moving to a service-centric IT model requires a transformational mindset. Transitional activities, albeit valuable in and of themselves, will not suffice. Both IT and the business leadership must recognize that a service-centric approach to IT better aligns and powers a service-centric business model, and thus requires serious attention to the substructure and very fabric of the IT organization.

To make it happen, both IT and the business entities must together focus on four key dimensions that constitute the service-centric IT transformation framework. We call these dimensions the four pillars:

1. Adopt the concept of the IT-provided service as IT's economic unit of measure. All IT deliverables must be organized around the notion of IT services being the economic unit of measure, the irreducible building block for supply, demand, quality and value. IT must think of its services as a product line, much like a manufacturer or retailer thinks about the products it offers to its consumers.

And in fact, this is the "service" in the term service-centric IT. Services—like payroll, sales forecasting, Blackberry, QA lab time-share, etc.—make up IT's product line. They are the things the business pays for, and from which value is derived.

Central to the notion of managing a service line, just like a product line, is a ecosystem for managing supply and demand in a way that's both fiscally responsible and serves the operating model of the company.

2. Orient IT to an enterprise architecture that supports the lifecycle of services and manages associated business value. If IT is going to function as a driver of innovation and business results, it must understand how it fits into the overall business structure, and create a business model for itself that fits in with that structure. Most IT shops use a governance framework called an enterprise architecture (EA) to understand the structure of the business as a whole. This is an effective strategy, but ironically these same IT shops don't have an enterprise architecture to describe their own structure. Clearly, there should be an enterprise architecture representation of the IT business model. We call this the EA for IT.

3. Use automation to realize IT efficiencies, exploit information and drive business decisions. With a structured operating model for IT in place, the service-centric approach lends itself more readily to automation across the entire organization. A clearly-articulated operating model drives greater understanding of how things are supposed to happen between silos, between tribes, and over a longer period of time. Furthermore, new approaches to automation (such as model-driven automation) and technical architectures (such as services oriented architecture, or SOA) support this type of structured, service-based orientation. Bringing this design paradigm into IT makes perfect sense, and in fact, leading IT automation providers are adopting this approach to make an ERP-like solution for IT more plausible.

This gets to the very heart of the notion of IT for IT—establishing how information management will be performed inside IT to support its strategic objectives.

4. Use an effective transformation methodology that delivers short-term results. Successful service-centric transformation is a matter of having the right plan and using a set of discrete, repeatable and measurable steps to make incremental progress toward the strategic end state. In later sections, we'll explore our recommended approach in detail, to illustrate how a service-centric IT organization emerges that is structurally aligned with the business model and institutionalized to provide lasting strategic value.

Figure 4–The four pillars of service-centric IT

Service-centric transformation requires moving away from the tribe mentality and toward a unified IT factory paradigm where every IT activity is an interdependent component of a value chain. To do this, IT must fully standardize both the services it offers and the tooling required to design, deliver and manage those services. In other words, it must enable a configure-to-order approach to replace the old build-to-order mentality.

Successful service-centric transformation is a matter of having the right plan and using a set of discrete, repeatable and measurable steps.

Ultimately, this transformation will result in improvements to the cost structure, speed and quality of the IT services delivered, which will satisfy the executive suite, the end users, and the IT organization itself.

Let's look at each of the four pillars in depth.

Pillar One: Establishing the service as IT's economic unit

The underlying principle of service-centric IT is that a discrete IT service is the basic economic unit of measure and exchange for supply, demand, quality and value.

Just like airlines provide transportation services and banks provide financial services, IT provides services as well. And just like other service-based industries, the service is the unit of exchange between the provider and the consumer.

All IT deliverables that get built, delivered, and managed by IT for the benefit of the lines of business are considered services. In the context of our discussion around enterprise IT throughout this book, the provider is IT, and the consumers are individual end users, business teams, other IT resources, or perhaps even end-customers of the business in a way that delivers measurable value.

The underlying principle of service-centric IT is that a discrete IT service is the basic economic unit of measure and exchange for supply, demand, quality and value.

Focusing on the service as IT's basic economic unit introduces a new organizing principle into the IT structure. To capitalize on this structure the IT organization should answer some very important questions:

- What is the portfolio of services that IT offers? What do they look like and how are they managed?
- Who are the consumers, suppliers, partners, providers? How do consumers acquire the services?

- What are the IT roles and functions required to support the service lifecycle?
- What are the common and independent processes that support each IT business function? How are the processes linked to ensure end-to-end visibility?
- What systems are needed to automate and adapt the processes?
- What information and metrics will be used to measure overall effectiveness of the model and quality of services?

Answering these questions is the first step in changing IT's mindset to that of a service provider. Once IT thinks differently, behavior changes will follow suit.

One of the major sources of tension between IT and the business is the lack of perceived value in the results that IT delivers. Moving to a service-centric IT model, services in these four areas can be built, packaged, delivered and managed more definitively and efficiently than as a silo-oriented collection of capabilities. In addition, these services can be linked with business outcomes in ways that make the value visible to both sides. Services can be measured against specific metrics for performance, availability, ongoing cost of ownership (for the lines of business), cost of maintenance (for the IT organization), risk and security, break/fix and return on investments.

So, just like a financial service or an airline service, the IT service is the focal point for the interaction between IT and the business. Creating the service as IT's central economic unit is the critical first step in transforming IT. By doing so, IT can change the nature of the conversation from IT cost to business value.

Pillar Two: Building an enterprise architecture for IT

For service-centric IT to work, IT must be organized and governed according to the specifications of an enterprise architecture. An EA is a comprehensive framework that explicitly defines all the relationships among the components of IT—technology, people, processes and information—in the context of the business as a whole.

An EA for IT does not currently exist in most IT shops, but not because IT executives think it is unnecessary. In fact, most IT organizations are working to create an actionable EA framing of their business. What keeps them from building one for IT is the common misconception that IT's energy should be spent in "helping the business be successful," as if investing in IT's operational performance wouldn't accomplish this. Without an EA for IT, the vision for how IT is supposed to work tends to remain locked in the head of the CIO. It might be brilliant, but no one else has access to it.

The EA for IT helps answer a lot of boardroom questions IT struggles with, such as: What value is IT delivering? Will this service-centric transformation really work? What does success actually look like? But that's only the high-level view. An EA for IT gives IT the leverage it needs to make service-centric IT work efficiently, down to the most basic level.

Capability-centric IT organizations are built around a form-follows-function model in which the structure of the IT organization itself is designed to fit the capabilities that IT offers. An EA for IT enables IT to switch to a function-follows-form model. Because IT's services and value chains are explicitly defined, the structure of IT flows from them, which makes execution more manageable and efficient, and results more consistent.

As an example of function-follows-form, consider the café latte at a franchise coffee shop. The franchise's objective is to deliver a latte that looks and tastes the same at all its 6,500+ locations. To accomplish this, it began with the end-product it wanted to deliver (the latte), and designed systems (a recipe, people, equipment, training systems, coffee bean delivery, etc.) that, when used in the proper sequence, result in that standard product. And like any business, it does this while understanding development costs, cost-of-goods-sold, and the supply chain required to ensure that all the right ingredients are in place when the consumer places the order.

Another benefit of the EA for IT is that it sets the structure and operating model of the IT function in the context of the structure and operating model of the business communities it services…assuming, of course, that the business models are indeed captured in the enterprise architecture framework. This helps drive greater efficiency across the enterprise by improving the alignment of IT and the business.

Finally, let's not forget the significant benefit an EA for IT brings to the people of IT. With the EA in place, IT employees better understand their role and how their work fits into the success of the business, which closes the "air gap" between employees and management teams, improves morale, and reduces turnover.

Before taking on the construction of an EA for IT, consider these questions:

- Do we clearly understand how to construct and implement an enterprise architecture framework?

- Are we willing to commit to governing IT using the enterprise architecture framework?
- How do we get the structure and operating model out of the head of the CIO and into a form where all IT personnel and the business can see and use it?
- How do we measure the results of change if we don't understand the as-is state of the IT function?
- What is the current organization design and where are the perceived or real bottlenecks?
- Since structure seemingly means bureaucracy, how do we avoid inserting roadblocks when implementing the new structure and operating model?

Formulating answers around these questions is critical to developing a sustainable EA initiative.

Before we complete this discussion, it's important to look at the hierarchy of the EA framework. The EA for IT is structured around four value chains: IT administrative management, business relationship management (BRM), service design, and service delivery.

It helps to think of these value chains using the analogy of one of the oldest service-centric business known to man: the restaurant. Clearly, the service that a restaurant delivers is the meal. The meal shows up on the menu, it has a price, description, configurable options (rare, medium or well-done), and an implied SLA (it is available to consume right now). The restaurant has very clear value chains: the owner and management staff comprises the administrative management chain. The wait staff and hostess are the business relationship management chain. The executive chef would represent the service design chain, and the line cooks, wait staff and bus boys comprise the service delivery chain. Each of these chains are supported by their representative operating models. There are processes for each of the chains, as well as information, automation and resources associated with each chain.

Every successful restaurant is driven by the interaction of the value chains, the support operating model, all around a very clearly defined economic unit: the meal. The is exactly how IT needs to function.

We'll go into these value chains in depth in later in this document. For now, it's important to acknowledge that the stability of the EA—and the service-centric IT organization as a whole—is dependent on the structure of the value chains and IT's ability to effectively position the service lifecycle across them.

Pillar Three: Automating for information-driven business decisions

In the service-centric IT organization, automation is used wherever possible to simplify, streamline and share processes and information. In this way, IT's services work better, and work better together to create maximum value and drive better business decisions.

IT automation, or IT for IT, can be grouped into three categories: task-based automation, process-based automation, and service-based automation.

Task-based automation is often embedded inside the technology delivered by IT suppliers. While it is important to making technology more useable and manageable, task-based automation also produces valuable information about the success or failure of specific units of work. This information is important to the service lifecycle, as we'll see later.

Process-based automation uses systems that automate and control workflow associated with the design, delivery or management of services to help reduce dependency on human intervention and reduce error rates. Process-based automation systems also are critical to improving the speed and predictability of service delivery. But these systems are only as good as the processes they are built around. In most capability-centric shops, a single process can have many derivatives, making process-based automation very complex and unwieldy. So the goal of process automation in service-centric IT is to build consistency into the processes, and by doing so simplify the automated controls.

Service-based automation provides controls that can be re-used across the ecosystem to promote integration and for a higher level of workflow control. This approach is a higher level of abstraction than process automation in that, much like SOA in the mainstream application space, it is comprised of components that can be called by applications and/or other services to complete work. While service-based automation is not new to most business application areas, it is a new frontier for the IT ecosystem. It is more applicable to a service-centric IT shop because of the model based view of the organization itself.

To ensure the success of automation initiatives, these are some key questions to be asked:

- Is there a pre-existing model for the design, deployment and management of the IT for IT?
- Do we have an inventory of all the automation currently in use and can it be mapped back to a role, function, process or other activity?
- What design paradigm should be applied to automate the IT work associated with the creation, management and delivery of IT services to the business?
- What is the systems architecture for IT?
- What is the information consumed and produced by the IT ecosystem?

Any type of automation to be implemented must be properly structured and connected with the EA for IT. Automation should be model-based and oriented to an overall service and data model, instead of a task or process. It should be integrated at the workflow and data layers in order to create predictable delivery cycle times and reduce dropped handoffs in the service pipeline. It should be built on an IT business structure, using the value chain as its highest level. Finally, automation should be information-oriented, providing the enterprise with maximum accessibility to the information being produced and/or consumed.

The idea of information orientation is not a new one for IT, but the execution is, because most of the organization's data is locked in tool islands and is accessible only to specific personnel who have access to the tool environments.

Implementing automation as part of an EA for IT, and *only* as part of an EA for IT, creates a well-defined and easily managed automated ecosystem for IT that is understandable to the business both in terms of its structure and of value delivered.

Used in this way, automation breaks down insular tribal structures and roadblocks caused by dedicated capability-centric tools. It also reduces the amount of human intervention required in the service pipeline, thus avoiding errors and keeping costs low. Automation can be a catalyst for process consistency and structural improvements, which has additional financial benefit. Using the EA for IT to build out the IT for IT helps move the organization from a collection of tools to an integrated systems architecture.

Pillar Four: Using a transformation methodology that delivers short-term results

The transformation methodology makes the concept of service-centric IT a reality. Almost every IT transformation initiative brings with it a level of anxiety, both for IT and the lines of business. Questions about cost, resources, and the disruption of business execution while the transformation is going on must be addressed prior to gaining acceptance and funding from executive management.

The transformation methodology we advocate helps IT lay out, in simple terms, how the activities will unfold, addressing questions such as:

- How will the transformation benefit the company as a whole?
- What methodology will be used? How will the methodology be implemented?
- Why will this approach work when previous attempts to improve IT's performance have failed?
- How long will it take to see results, and what tradeoffs will need to be made?
- How will IT continue to function during the transformation?
- What benefits can be expected as the process moves forward?
- What does success for this transformation look like?

Part of the reason why service-centric transformation will be viewed with such skepticism is that, for most companies, it isn't the first time IT has tried to transform itself. In the past, IT hasn't made it across the finish line.

The failure of other types of transformation initiatives can be attributed to several major factors. First and foremost, there is the "big rock" syndrome: because of the nested complexity of most IT environments, plans to attack problems quickly become larger than originally scoped. Think of these as the big rocks. Big rocks are heavy, slow to move, require many resources to get going and are quite disruptive once they are moving. So tackling big rock initiatives almost always results in a loss of momentum and focus, which over time turns into project atrophy.

IT organizations often undertake the process of transformation without clearly understanding the underlying details associated with achieving the end state: the transformed organization (described earlier). This creates tension at almost every level—especially when IT can't show short-term results, and there are interruptions in business execution.

This inability to deliver short-term benefits is another cause of transformation failure. The business is used to investing its money in things that provide visible and measurable returns over a short time period. Because typical transformation approaches have a long tail, returns don't appear fast enough, and the boardroom loses patience and decides to cut its losses even though the transformation itself may actually be working. In cases where cost savings do materialize, they often are usurped to improve quarterly financial results instead of being re-invested in the transformation process.

Finally, personnel issues often derail transformation. As we've discussed, the IT organization operates continually in an understaffed, oversubscribed state. Most IT employees are 100% occupied with pre-existing work loads. To transform IT, tradeoffs must be made to free up staff to work on new initiatives. Very often, these tradeoffs are too difficult to make, and transformation initiatives move down the priority list until they ultimately disappear. In addition, transformation requires unique skill sets which are already in high demand and are the most difficult to reassign. The time-to-productivity for hiring external resources to drive transformation initiatives, or replace those existing who can be reassigned, often exceeds the tolerance threshold for the initiatives. Finally, changing business priorities wreak havoc on IT staffing and personnel decisions. Because keeping the business execution on track is the number one priority of IT, these personnel issues often stifle the progress of transformation efforts.

What we propose is a streamlined transformation model that accounts for and helps avoid these major pitfalls associated with typical transformation projects.

By using a simple, iterative transformation methodology, the IT organization can successfully bring about service-centric transformation and maintain that success over time. With the EA for IT as a guide, and by creating a value chain based model, IT can focus on transforming a few services at a time .

Keeping the scope small helps IT avoid "big rock" atrophy, maintain momentum and show quick success. This scale of change also keeps disruptions and complaints to a minimum, and produces significant short-term benefits.

Using a land-and-expand approach, additional services can be identified and transformed once the original set of services has been completed. A transformation snowball effect is eventually created as artifacts and techniques from previous service transformation iterations are re-used.

Because this step-by-step transformation methodology creates a continuing series of incremental successes, instead of one huge success that only happens at the end, service-centric transformation can survive a change at the CIO level or among other IT staff, a change in funding, and changes in business priorities. Course corrections are simple to make. Value created from early transformations can be reinvested to fund future ones, and even if it becomes necessary to shift priorities while transforming a particular service or set of services, it is easy to close out the process and come back to it later.

Creating forward momentum toward success

Success is the goal of any initiative, but in the case of service-centric transformation it is critical to maintaining momentum, focus and business support. By maintaining focus in these areas, IT will create clearly visible successes:

- A well-defined service taxonomy and a catalog of services which are configurable and include accurate price quotes and delivery time estimates
- An Enterprise Architecture for IT that reflects the actual structure and operating model for how work gets done
- An operating model for IT with clear integration and information models
- An ever-growing number of transformed services, and in the productivity, efficiency and financial benefits delivered by these services

With these successes building upon each other, the forward progress of service-centric transformation will be very hard to stop.

A Foundation for Lasting Benefits

Maintaining success over time is vital to service-centric transformation. Transformation initiatives can yield both short-term and long-term benefits, but without changes in IT fundamentals, these benefits will not last. It's like going on a diet until your target weight is reached, and then eating again in the ways that caused you to become overweight in the first place. Service Centric transformation—changing IT from a capability-centric model to a service-centric line of business—addresses IT issues at the root cause, and thus creates lasting benefits in four areas:

- **Agility:** The transformed IT organization is the underlying support system for every employee, every process, every transaction in the business. IT has the flexibility to adapt quickly and easily to the inevitable changes that happen in the business.
- **Economy:** By structuring IT around the economic unit of service, maximum value is extracted from each project or product, which enables IT at last to "run like a business."
- **Control:** With the EA for IT and value chains in place, IT can better manage its portfolios of projects, assets, suppliers, people and services, thus reducing the risk associated with each new project.
- **Alignment:** Executive management and the businesses understand both how to engage IT to get results. They understand how IT delivers value to the business. IT is fully able to perform its proper function in powering business success.

The goal of service-centric transformation is to link IT and the business at a single point of control—the point of service—and to empower IT to create the right fundamental structure to support the ongoing management of the lifecycle of each service. To achieve this, IT and the business must strike a balance in four areas:

Cost: IT must balance the spending on transformation with changes to its cost structure, reducing spending wherever possible and providing greater cost transparency and visibility to executive management. In addition, IT must improve the utilization of all its assets (servers, storage, networks, facilities, people, etc.)

Speed or flexibility: Specific improvements must be made with the end user in mind, always focusing on making it easier to do business both internally and externally.

Quality: IT must make changes that improve both performance and organizational effectiveness, such as quality of delivery, quality of information, and quality of staff.

Risk mitigation: The transformed IT organization must support profitable business growth, deliver business value instead of cost, produce predictable outcomes, enable regulatory compliance and minimize security exposure while at the same time helping the enterprise capitalize on IT as a competitive weapon.

Changing IT from a capability-centric model to a service-centric line of business addresses IT issues at the root cause, and thus creates lasting benefits.

Making it happen

We began this section with a very big question: with so much at stake and so much invested in the status quo, how can service-centric transformation possibly be a good idea?

The answer to that question is two-fold. First, the status quo is unworkable and unsustainable. Second, if the IT organization is to have any hope of long-term success, it must abandon the status quo and reinvent itself from the ground up around a structured model.

That's the bad news.

The good news is that service-centric transformation offers IT a pragmatic, workable and sustainable solution. It provides the opportunity to achieve the kind of DNA-level change IT has always wanted to make, but could never fully accomplish. It is a simple, step-by-step approach that keeps the focus on small, incremental transformations instead of big, complex ones. It delivers short-term benefits that even your CFO will love, and it can survive the personnel changes and shifting business priorities that will certainly come along the way.

So service-centric transformation is a good idea not only because it eliminates the worn-out capability-centric IT model, but also because it aligns IT with emerging market trends. As we said at the beginning, it is a complex, large-scale, time-intensive initiative that involves upending the status quo and changing the way IT operates at the most basic level.

But it does succeed. It is worth doing. And now that you've made the decision to do it, we can look more closely at how service-centric IT transformation works.

Pillar One: The IT Service

Just Exactly What is a Service?

Service is a word that has many different definitions. So let's define it clearly: IT services are the *products* of IT—discrete units of consumable functionality consisting of technologies, products, process and/or support that together comprise value to a consumer. IT can assign a value to each service, which is "sold" on a transaction basis to lines of business, end users and even other groups within IT.

The concept of IT services is very familiar to IT professionals—in fact, there are in most IT organizations a fair number of service level agreements (SLAs) designed to measure the performance of IT against stated business goals or outcomes. But most SLAs actually are oriented toward IT capabilities, such as delivering 99.999% network uptime or sub-second application response, rather than actual business-oriented services. In addition, most SLAs aren't attached to real and specific business value that is either delivered to or derived by customers.

IT services are economic units—in other words, the requester is willing to purchase and pay for them as discrete offerings that carry implicit value. Therefore, the parameters of each service (who is entitled to get it, how it is requested, what it entails, what business outcome results, how much it costs) must be clearly defined. Otherwise, the customer doesn't fully understand what he or she is buying.

Just like at an online bookstore or the supermarket, transactions of IT services are enabled through a series of provider-consumer relationships which flow through various value chains. A product is designed and created and then delivered via a marketplace, a consumer decides to purchase it, money is exchanged, and value is delivered in the form of that product. Underlying those relationships, of course, there is a management function that keeps transactions between IT and its consumers running smoothly at every point within the various value chains.

Service Providers and Service Consumers

Before we begin looking at these relationships, it's important to recognize that every participant in the value chains of any business operate *both* as consumers and a providers. Figure 5 illustrates this. In the provider role, an entity consumes services that allow it to operate, and as a result of this operation, creates services or products that it delivers to its customers in the consumer role.

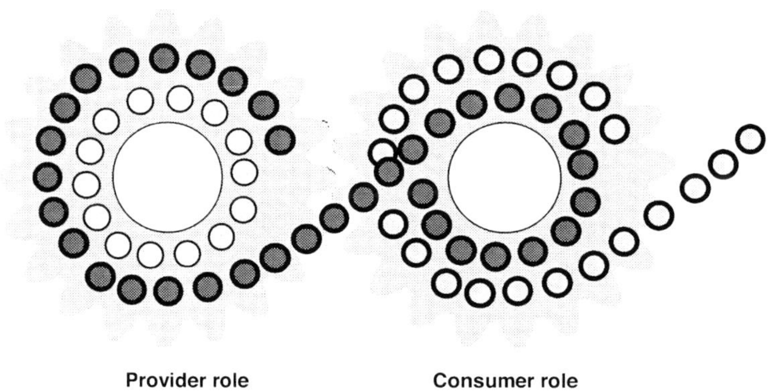

Figure 5

The customer buys the services or products and consumes them to enable its own capabilities as a service provider. This simple dynamic repeats itself until the resulting service (or product) reaches the end-customers served by the business.

Now let's look at these provider/consumer relationships in the context of the IT organization (Figure 6). In the provider role, the IT organization consumes products and services such as service desk technologies, change management systems, blade servers, facility services, network circuits, employees and legal services. Think of these as *IT-consumed services*, or the "tooling" of IT, which enable it to operate. The dots then represent these IT-consumed services that are used inside the IT factory. As a result of these factory operations, *IT-provided services* are created for use by the business. These might include consumer credit checking, employee expense reporting, new employee set-up, or a private network.

Offering these IT-provided services through something as simple as the service catalog exposes the providers' offerings to the consumer role. These consumers procure the IT-provided services to enable their business services.

As before, the cycle repeats until the desired products or services reaches the company's end-customers.

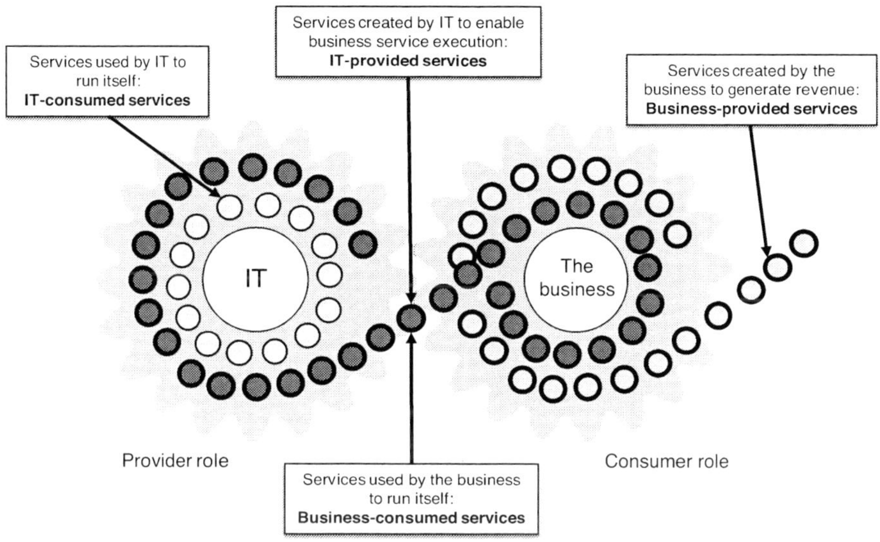

Figure 6

What these IT-provided services have in common is that they become an economic unit of measure to the business. In other words, they are designed, constructed and delivered in such a way that the consumer (the lines of business) derives value from them, and is willing and able to purchase them.

In fact, IT-provided services transform into to business-consumed services at the moment of payment. This financial exchange is indicative of the value as perceived by the consumer.

Note that, by their very nature, IT-provided services—not the process, the people, the technology, or the applications alone—make IT a service provider. These discrete services, defined as the economic unit of exchange, are key to enabling IT's transformation from a capability provider to a service provider.

A motivating force for adopting a service-centric model for IT is that IT-provided services can be linked directly to value-based business outcomes, and governed by a service level agreement. Such a model also creates a mindset focused on managing a portfolio of services, or a service line (just like a consumer goods product line).

Now let's extend this paradigm further, to the business as a whole.

As you can see in Figure 7, the business itself is a macrocosm of the provider/consumer relationship. Business is all about consuming and assembling services and products from multiple providers in order to create and offer goods and services to customers. Banks combine policy, intellectual capital and technology to deliver financial services; coffee shops bring together roasted beans, water and espresso machines to deliver lattes; airlines use planes, baggage carts and microscopic bags of peanuts to deliver travel services. From the end-customer's point of view, the business deliverable (cash from an ATM, the cup of coffee, the airline ticket) is the economic unit of exchange, or the business-provided service.

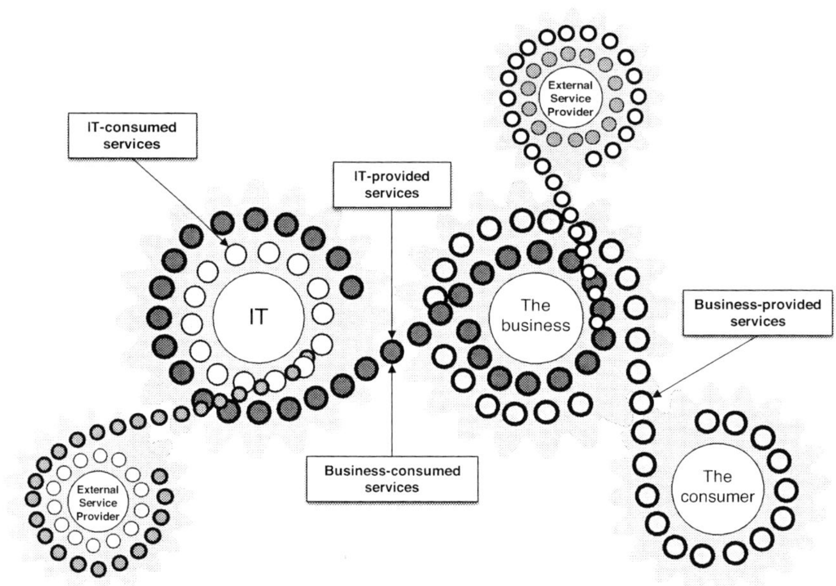

Figure 7

For any service business, the definition, pricing and valuation of business-delivered service is the single most important organizing principle.

The health and success of the business depends upon doing that job well, year in and year out. So if IT models itself according to the principle of "service as economic unit" and sees itself as an integral part of the larger service ecosystem, it can begin to structure its operations based on value chain dynamics.

> If IT models itself according to the principle of "service as economic unit" and sees itself as an integral part of the larger service ecosystem, it can begin to structure its operations based on value chain dynamics.

Structuring IT Around Services

Managing IT-provided services as units of economic exchange means managing the cost of goods sold while investing in a business structure that will continually produce a product line that is stable, profitable, and delivers value to customers.

To drive efficiency and scale, IT must be able to establish and maintain the proper ratio between build-to-order and configure-to-order services (approximately 80:20% in favor of configure-to-order). An example of a build-to-order IT-provided service would be the development of a mobile messaging application using a set of programmers writing code to meet a certain set of specifications. Using tools and techniques that the programmers are comfortable with, the end results calls for unique server, network and security configurations, which in turn results in a data center albatross that is expensive to maintain. As every IT executive knows, the build-to-order approach to service delivery is generally time-consuming, resource-intensive, expensive and has an unpredictable delivery cycle.

In contrast, a configure-to-order IT-provided service might be a PDA messaging service, where all the solution elements (the device, service from a vendor, and support) are based on pre-defined models and require little time, effort or expense to produce and maintain the desired outcome for the consumer.

So the structure of the IT organization must be built on two foundational ideas:

- **The service deliverable**, which is designed for both usability and manageability. Services become increasingly financially efficient (profitable) when the cost of goods sold is effectively managed. Thus the need to maintain that 80-20 ratio between configure-to-order and build-to-order services becomes evident.
- **A business model** for the IT function that is sufficiently funded and organized to support the continued improvements required for the production, delivery, and management of services. Like any successful

business, IT cannot function properly without a clear and well articulated, structured model.

With this structure in mind, the next step is to build the services IT will offer to its customers. This starts with a methodology for defining services in a way everyone involved can understand.

Defining IT-Provided Services

To define and then design each service offering, it's important to understand the major criteria that should be used to help shape them. These include:

What customers are we serving? The economic buyers of the services should be identified and profiles developed for each type or class. For example, there might be a profile for engineers that differs from manufacturing or administrative types of service consumers. While the consumer of IT-provided services typically lives in an internal business unit, in some industries (such as telecommunications and investment banking), the external customers of the company are the actual consumers. Furthermore, the economic buyers of IT-consumed services reside in the IT organization. Services like network and system management are rendered to other IT professionals. These different profiles are important because just as the need exists for changes in behaviors within IT, the economic buyers of IT services will need to change as well—for example they will now be ordering services instead of servers. What these services cost, how to obtain them and what to expect from them will require a certain level of retraining. Therefore, as the economic buyers of the services are identified and profiled, designs, organization structure and education plans can be developed around them.

What value can we deliver to these customers? Knowing the customer is only the beginning. IT must also match its services with the needs of its customers in order to create offerings that deliver real value, and for which customers are willing to pay. Framing services around value requires a rich understanding of the company's business goals and needs over a specified time horizon. It's important to separate principles from goals here. For example, a principle might be profitability, while a goal might be for the company to move to a direct sales model from a channel sales model over the coming year. Understanding the goals helps IT design and position its services in the context of the value expectations of the customers.

What roles and relationships are required for the service lifecycle? To complete the definition, IT must also identify the people and processes responsible for creating and maintaining the service, along with the specifics of how they interact. This includes establishing ownership of the service line and the service offering itself, the roles required to design, build, deliver and manage the service over the course of its lifecycle, the metrics to be used to measure the service, and the different ways the service can be bundled with other services to create value. Service-line ownership roles should be established and metrics defined, much like how product-line ownership is managed inside the businesses.

Defining the structure for the service line is a unifying force that aligns IT services with the business services they enable and helps IT measure and manage performance against meaningful metrics. In addition, managing the service lines as a portfolio enables overall improvement to the investment strategy within IT and the business.

Armed with these facts, consider the definition process. Every service needs:

- A name: for example, "Sales Forecasting Service"
- A description: "A daily worldwide sales pipeline report and analysis for senior sales management"
- An expected business outcome: "Providing accurate sales projections to guide executive management and ensure top-line transparency for the financial analyst community."
- A complete set of artifacts and descriptors: These might include the service level agreements that align IT with the business around performance and availability, a logical and physical view of the configuration items that underpin the service, a view of dependant services, documentation, a continuity plan, knowledge entries, subscriber entitlements, and security and access provisions.

Service definitions, like the services themselves, do not exist in a vacuum. They must relate and integrate with other service definitions and the IT organization model as a whole. This requires the discipline to maintain not only a consistency among service definitions, but also in IT's approach to developing those definitions.

The process of defining the services IT will offer its customers is virtually the same process used by any business to develop a model of its service or product lines. Thus, the service-centric model for IT and the idea of "running IT like a business" are essentially interlocked.

As more and more services are defined, it will be important to collect them into logical groupings for the sake of manageability and usability. Most IT-provided services can be grouped into four basic categories:

Employee services (sometimes called productivity services) are delivered on an individual basis to help people work faster and more productively.

Application services are delivered to the business workgroups or functions collectively to enable processes and information exchange.

Information services are delivered to individuals and workgroups to support functions and processes. They provide decision support, customer visibility or process integration.

Infrastructure services are delivered to the enterprise to enable other services, or delivered to IT workgroups to improve the productivity of their business communities.

	Description	**Example**
Employee services	Delivered on an individual basis to enable worker/office productivity	cell phone, pager, email accounts
Application services	Delivered to the business workgroup or function to enable processes and/or information exchange	application development, hosting, EDI
Information services	Delivered to individuals and workgroups to support functions and processes. Provide decision support, customer visibility or process integration	sales order forecast, data warehouse, data models
Infrastructure services	Delivered to the enterprise to enable other services or delivered to IT workgroups to enable productivity of their business communities	network, servers, call center telephony

Figure 8 - Example service categories and descriptions

Individual service offerings exist in each category, and the names will vary from company to company. But the deliverables of virtually every corporate IT shop today can be categorized this way. This is important because using these categories enables the creation of a common taxonomy, which in turn drives improvements in structure, measurement and investment decisions.

How Services Create Value

We can see how the *service* can become the basic economic unit of IT—the "products" IT offers for sale to its customers, aka the organization's business lines. By understanding the cyclical nature of the provider/consumer relationships involved, IT-provided services can create value by enabling consumers to build and provide services and products to their own customers. IT can structure itself around these services, concentrating on maximizing the number of configure-to-order services offered, and minimizing the build-to-order services in its portfolio. This structure is best expressed as the operating model for IT, a crucial element in the process of service-oriented IT transformation—which we will get to momentarily.

To be effective, IT-provided services must be properly defined so that both the producers and consumers of the service have a consistent understanding of the associated business outcome and derived value. Consistency in the definition process and in the definitions themselves will help IT manage its service portfolio more effectively and keep services relevant and valuable over time.

Finally, categorizing IT's service line into employee, application, information and infrastructure services helps create a common taxonomy that minimizes confusion and helps drive better measurement and management decision making.

It is also imperative to formulate the blueprint for building a service-oriented organization through the enterprise architecture for IT.

Pillar Two: The Enterprise Architecture for IT

An Organizing Framework for IT

The second pillar of service-centric transformation is orienting IT to a comprehensive enterprise architecture. With services established as the product line, IT can be structured in such a way that the economics, performance and the delivery pipeline for those services can be managed and measured effectively.

The EA for IT accomplishes this. It is an organizing principle for business process and IT capabilities reflecting the integration and standardization opportunities in the enterprise's operating model. Before going into detail about the EA for IT, we should first look at how enterprise architectures work for the business as a whole.

At the enterprise level, the EA is a comprehensive framework that defines relationships among the components of the business, and provides a context for ongoing execution of changes within business functions. The enterprise architecture is generally represented by a core diagram that encapsulates business value chains, associated processes, shared data driving the processes, key automation technologies and customer constituencies. It typically is created in one of two ways: as the result of conversations among the business leaders then shared with IT, or as a design created by IT, and reviewed and finalized with the business leaders. Either way, the EA itself serves as the representative embodiment of the corporate strategy and becomes the context for the architecture of each business unit.

Every line of business within a company is structured using either an explicit or implied business unit architecture—and IT typically tailors its model to meet the parameters the architecture sets forth. The business unit architecture captures the details of how the line of business operates, and serves as a foundation for execution. A typical business is organized around four basic value chains:

- Business development chain: generating sales and revenue for the line of business
- Customer chain: delivering products, meeting customer needs and handling customer issues
- Design chain: creating and updating products for the business line to sell
- Supply chain: gathering the necessary components and building the business line's products

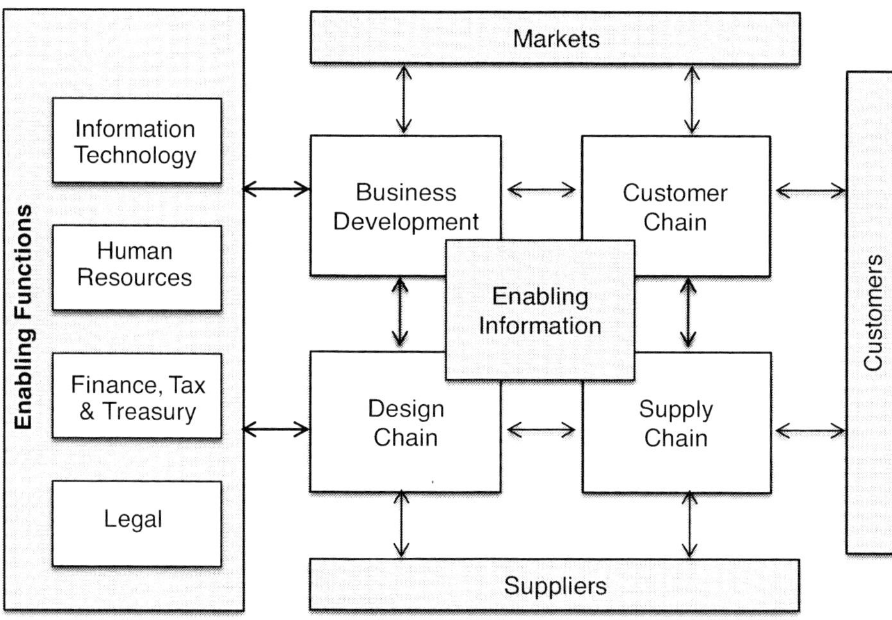

Each of these value chains interacts with the other three, and all are supported by enabling information and functions. Externally, the architecture reflects interactions with customers, suppliers, and markets to shape and extend overall behaviors and capability of the goods and services delivered. In this way, the EA frames the necessary level of process, information, resource and automation required to deliver goods and services to customers.

Enterprise architecture supports the definition of functions and roles

A business, or a line of business delivers some physical product or service to customers—call these business-provided services, or just business services. These business services may be a product that enables the customer to perform some act, or a service or product that performs an act on behalf of or for the customer. The core diagram in the EA depicts the mechanics behind these functions along with their interactions. Understanding the functions and flow help businesses better position themselves when making sourcing decisions, such as what should be retained, and what should be outsourced.

The automobile is an example of this idea. General Motors manufactures automobiles for customers. The tangible business services involved are the production and distribution of automobiles. The soft business services are things like financing, insurance claims services, On-Star service and general repair services.

Each function called out in the EA becomes comprise people, process and technology, and fill specific roles that make the services real to customers. Understanding the natural strata of EA functions help businesses construct strong organization structures. A strong organization structure helps employees understand where they fit in and how their deliverables contribute to the overall chain of delivering goods and services to customers.

Enterprise architecture helps measure results

Another key function of the enterprise architecture is that it provides context for how results should be captured, measured and rewarded. It provides a way to measure the performance of the line of business it represents. Business services are typically measured across four vectors:

Financial: How much does it cost to deliver and operate the business service? How can the value to consumers be measured?

Quality: How good is the business service? Will customers buy it?

Risk: How much risk is tolerable in building and delivering the business service?

Agility: How quickly can the business service be brought to market and be adjusted to changing conditions? How can new customers in new markets best be reached? How can new labor markets be used to build the product? How easy is it to adjust to changing customer needs and market conditions?

Of course, each business service is different, so there is no one definition of success in terms of the relationships of these four measurement vectors. But, as the saying goes, without measurement there is no management. So together, the four measurement vectors allow the business line to review and validate progress toward meeting its overall objectives.

Applying Enterprise Architecture Principles to IT

For many years, IT has used the enterprise-level EA to model the characteristics of the businesses they support. But until now, IT hasn't really had the time or motivation to apply the EA framing to its own operations in such a way that accurately depicts the IT business function. While other lines of business were developing and maturing their business unit architectures, IT was maturing and managing itself by implied models and learned behavior. That strategy was good enough to get by in small or low-growth environments, but in large enterprises this approach is simply inadequate.

In fact, the lack of an EA for IT is one of the key reasons for IT's long-term inability to transform. If you can't articulate where you are, you can't know where you're going, or how you're going to get there.

The EA for IT is the business unit architecture for the IT line of business. It's made up of four value chains and an operating model that has four domains and a group of sub-domains. It provides explicit guidelines that serve as an anchor point for initiatives that create measurable advancement in maturity—in effect, becoming the foundation for service-centric transformation.

The EA for IT helps streamline the organization structure, identifying and eliminating redundant activities and connecting the organization across all levels. The EA for IT also provides the lines of business with the visibility they need to engage with IT, understand the value IT delivers, and get the outcomes they need to meet the expectations of the end-customers.

The enterprise architecture for IT is best represented as a two-layered structure, which fully models how IT functions in the role of delivering IT-provided services to the enterprise.

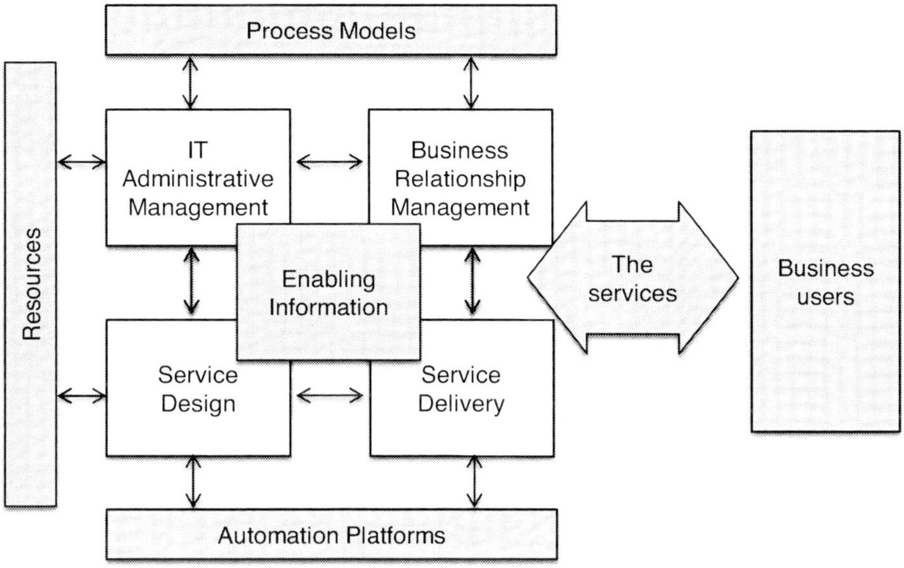

Figure 10 - The enterprise architecture for IT

IT's four value chains

The first layer is the value chains, which are the highest level organizing unit for categorizing the work in the IT line of business. Remember that a typical line of business has four value chains: the business development chain, the customer chain, the design chain and the supply chain. IT also has four value chains, which closer mirror those of the business:

- IT administrative management chain (akin to the business development chain)
- IT business relationship management chain (akin to the customer chain)
- IT service design chain (akin to the design chain)
- IT service delivery chain (akin to the supply chain)

We'll discuss IT value chains in detail later. For now, it's important to understand the relationship between IT services and the value chain layer of the EA for IT. Service requests are projects or individual transactions that originate from a line of business or an end user. In either case, some entity is buying an IT service. Project or service requests typically originate through the IT business relationship management value chain, while individual service requests are typically rendered through the service delivery chain. Each value chain is equipped to handle the order, making it easy for IT to manage the service lifecycle.

The second layer of the EA for IT is the operating model, which depicts the actual mechanics of how the work is performed—in effect, how IT functions as a business. It comprises 4 layers: process, automation, people and information. And, it describes the core and supporting functions IT provides, the work breakdown structure within each IT value chain, and all the inputs and relationships associated with the business, with suppliers, and with external partners. From the value chain categories above, a service request passes into the IT operating model layers, where it is handled via the operating model.

A Closer Look at IT Value Chains

IT value chains specify the categories of IT work and add a level of structure that serves as the context for all process, automation and people roles across the organization. Value chains are at the very core of the EA for IT, because they provide the most effective way to define and categorize how IT works, and thus establish a context for simplifying the organization and/or automation of that work.

By adopting the value chain model, IT is able to quickly improve consistency, transparency and reliability, and take a huge step toward achieving the transformation objective of making service the fundamental building block of every IT deliverable. Value chains enable IT to better manage supply, demand, cost and quality across the enterprise, while providing an anchor for processes, information, automation and resources. They clarify the division of labor, providing clean lines of demarcation and integration points for each task. Value chains also enable IT to create a system of record for a given function or task, and use the data associated with that task to create a single version of truth that serves all stakeholders.

For example, think about the provisioning of network bandwidth. Telecommunication costs eat up one-third of the IT budget in many large corporations, and a major part of that is payments to carriers for the necessary network bandwidth. When the carrier bill arrives, usually for millions of dollars, a reconciliation is generally needed to iron out differences between what the carrier says was used and what the company's internal IT tools reflect. With different sources of data all seemingly right, this reconciliation process quickly becomes tedious, time consuming and error prone—so much so that some companies actually outsource the job. But with so much money at stake, the job must be done, and done again every time a new bill arrives.

Defining this function in the context of a value chain eliminates the problem. The value chain deconstructs the network bandwidth function into a series of discrete units of work that form a service, and provides a single, authoritative source for collecting, managing and analyzing the operational data that result from the performance of that service. When the carrier bill arrives, no reconciliation is necessary because both IT and the carrier are using the same data.

Value chains work the same way for financial data, HR data, and any other kind of data used by IT. What results is an ERP-like structure for IT, where automation is used to facilitate and improve the access to and delivery of IT services.

How IT value chains work

The **IT administrative management** value chain orchestrates the running of IT business functions including finance, HR and legal, as well as portfolio management, strategy and governance. This value chain creates continuous improvements in both the services of IT and on the performance of the IT organization itself, i.e. how and how well those products are delivered. Roles within the value chain include the CIO, PMO, portfolio managers, finance, contract management, procurement and HR. The IT business management chain takes a horizontal view across the business in the planning, prioritization and funding of work requested of IT. This means that as demand signals roll up from a particular business unit in the form of project requests, they are prioritized against other demands from across other business units to attain a single IT plan of record. This action places the burden of economic business value definition back on the business leaders, where it rightfully belongs. Any disagreements on IT investment priorities can be addressed in the boardroom instead of the computer room. This value chain also promotes a controlled evaluation and investment re-balancing to ensure business demand and IT supply capabilities are aligned.

The **IT business relationship management** (BRM) value chain is focused on capturing and managing demand for new or enhanced services, aligning IT innovation with the business strategy, managing the service portfolio for a particular business unit, along with SLA management. This value chain includes engagement with the various lines of business, and the measurement of business value for all IT-provided services offered to those lines of business. A typical model for the BRM chain would be the assignment of a team for each line of business, responsible for capturing demand for IT services and managing service level agreements. Roles within the value chain include project managers, application implementation teams, systems architects and relationship managers among others.

By moving through the IT BRM value chain, demands from the business are consolidated, rationalized and prioritized for maximum business value. New and changed services to be offered to the end user community are assessed and prioritized, services and projects are properly resourced, and the relevant needs of finance, legal, HR and other supporting department are gathered, analyzed and met.

The **service design** value chain focuses on designing, configuring or building and releasing IT services that match business requirements with IT capabilities. This includes service lifecycle management, design and development, sourcing and assembly of components, testing of new and modified services to ensure quality, and transitioning services from design to implementation and ongoing service delivery. Key roles within the value chain include IT architects, service managers, network and systems engineers, application developers, database administrators and security specialists among others. By moving services through the service design value chain, IT ensures that its offerings meet the needs of its business partners, are designed for optimum supportability, work as advertised, and deliver the best return on investment.

The service delivery chain focuses on fulfillment of service requests from the user community, operational implementation support for services driven through projects, and the ongoing work associated with running the IT-provided services. This includes service operations, service sourcing and fulfillment, operational level agreements (OLA) management, and management of outsourcing partner relationships, as well as mitigating and repairing service disruptions. Key roles within this value chain are typically the IT operations staff, support engineers, service delivery specialists and escalation managers. The service delivery chain keeps IT in constant communication with both its sourcing partners and the lines of business it serves. This makes it easy for IT to monitor the implementation and performance of its services in everyday use.

These four value chains are interdependent. In the transformed service-centric IT organization, a request for service can pass through each value chain. Although the value chains offer a foundational structure that does not change, the work performed within each value chain may vary based on the operating model design specific to the individual IT organization positioning in the enterprise.

Expressing value chains as lifecycles

As the word "chain" suggests, the value chain is, at its heart, the movement of a service along a continuum. Thus, each value chain is best expressed as a lifecycle, providing a structure for governing the work associated with each service. These lifecycles describe all the high-level phases of work for each aspect of providing services to the business, and all lifecycles are built upon a foundation of shared data and information. Because it is a lifecycle, each has a specific beginning, middle and end. The nomenclature used here is meant to provide guidance to the reader for what a value chain lifecycle should look like. Modifications of the lifecycle phases may be necessary depending on the level of maturity inside the IT organization itself.

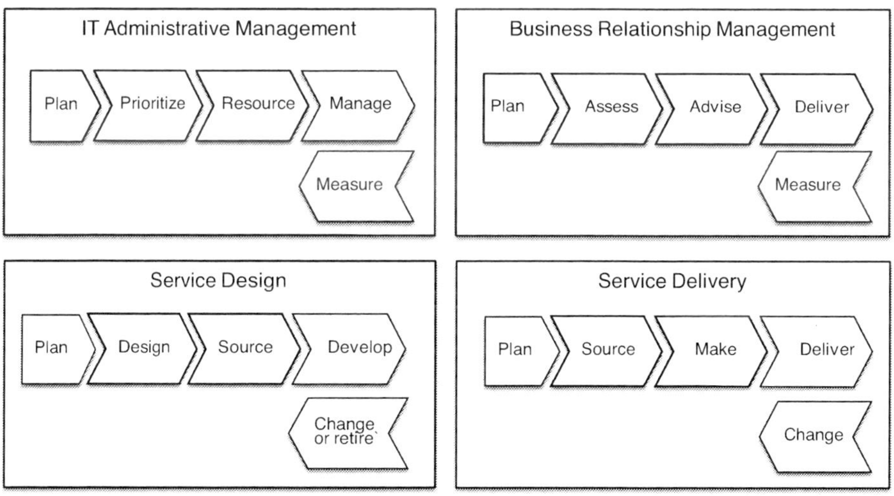

Figure 11 - Value chains are expressed as a lifecycle

Simply said, the work of IT is organized into four value chains, and each value chain is expressed as a lifecycle. In turn, each lifecycle is made up of phases. Next, we'll explore the phases of each value chain.

The lifecycle of the IT administrative management value chain

The lifecycle of the IT administrative management chain is primarily about managing the horizontal project portfolio and ensuring the alignment of business demands with IT supply. In a sizable enterprise this is no easy task, since there are many moving parts that must be orchestrated efficiently. The lifecycle proceeds through these phases:

Plan: This phase begins with developing, publishing and managing the plan of record and adjustments to the master portfolios for the coming period. It requires visibility into data from the demand side (the BRM chain) and from the supply side (the design and delivery chains), along with executive-level sponsorship in each business unit. Each project request must be evaluated for business value and aligned with supply-side cost and capability. Questions to be considered include: How much will the project cost? Do we have the technical capacity? Do we have agreements in place with suppliers for licenses and/or volume purchase agreements? The end result of this phase is a draft plan of record or project portfolio for the upcoming period.

Prioritize: The draft project portfolio then is reviewed with business leaders to set levels of expected investments and business benefits. In this phase, it is incumbent upon the business leaders (rather than IT) to establish priorities, and for these leaders to agree to and support the stated business value of the approved requests. Once this phase is completed, the official project portfolio is in place.

Resource: This phase involves the assignment of human and financial resource to perform the work specified in the project portfolio. IT must ensure that the appropriate skills are made available or are obtained, along with executing any necessary contracts or outsourcing activities that might be required. Resourcing can have long lead times depending upon the availability of the appropriate skill sets, technologies and suppliers for the work. In addition, any additional necessary funding sources, such as capital allocation, must be obtained during this phase. The outcome of this phase is generally the "go" signal to project teams to begin work.

Manage: With work now ongoing, this phase focuses on administering the execution of the project portfolio against stated objectives. Reports from project teams and other sources are assembled and output as status updates. Any escalations, roadblocks or other issues are handled in this phase, which runs until the project portfolio is completed.

Measure: Upon completion of the project portfolio execution, the results are analyzed against predetermined benchmarks to ensure that the expected benefits were realized, and if not, why not. This knowledge is used to foster continuous improvements that are applied to the next iteration of the lifecycle.

The lifecycle of the BRM value chain

The BRM value chain ensures ongoing alignment between IT and its business consumers. This chain typically focuses vertically within a particular line of business so that requirements, demand for new services and ongoing service level management can occur in the context of the structure and operating model of the business. Phases of the lifecycle are as follows:

Plan: The lifecycle begins by gathering demand signals and requirements for new or existing services. Some demand signals are captured as project requests, while others are projections of increases (or decreases) in capacity needs for existing services. Performing demand management, financial modeling and resource planning should be done in conjunction with changes or shifts in the business goals.

Assess: Next, the demand signals are evaluated and cost-benefit analysis is undertaken. This requires a keen understanding of the nature of requirements, business opportunities, market differentiation, and competitive parity opportunities. This phase is critical to capturing a clear view of the economics surrounding a particular request for work. The output of this phase is used for horizontal prioritization within IT.

Advise: This phase begins the development of project plans with the business. New innovation opportunities, technical architecture, and conceptual models for services are brought forward.

Deliver: With plans in place, execution begins. This likely will include the service design chain. Deliverables are configured, tested, and moved forward to implementation.

Measure: When the plan is fully executed, results are measured. Also in this phase, opportunities for continuous improvement are identified and presented to IT and the business. IT performance against stated goals is measured in the form of an SLA.

The lifecycle of the service design value chain

The service design value chain is where most of the project work that fulfills demand for IT services is performed. The output of the Service Design chain is a service that is ready for purchase and can be handed off to the service delivery chain for ongoing management. Independent service offerings are bundled together or configured and moved forward through a development, testing and deployment cycle. When requested services requires engineering work in order to be completed, such as a build-to-order offering, the appropriate resources are assigned to perform this work. In addition, activities associated with the ongoing service management lifecycle are in play including service costing, modifications, supplier relationship management. The lifecycle is made up of these phases:

Plan: Planning in the service design chain happens on several levels. First, there must be structured activities to develop plans for new services or modifications to existing services—usually an annual process. There also is planning that occurs at the project level, based on the project portfolio ratified by both business and IT leaders. This planning generally occurs in conjunction with the project portfolio activities resulting from the BRM chain. Finally, there must be planning at an architectural level to ensure the structural integrity of IT-provided services. This activity happens at least quarterly during the year. As projects complete and service changes are brought in, the architecture review board must meet to ensure the integrity of the IT architecture.

Design: Next, the conceptual models of the planning process are turned into physical models—artifacts such as renderings, data models and flow diagrams, application architecture renderings, infrastructure diagrams, and database design schemas.

Develop: Services are sourced, configured, developed and tested in an environment that closely mirrors how the service will perform upon implementation.

Implement: The service is transitioned into production, working in tandem with resources from the BRM and service delivery chains.

Change: A continuous stream of feedback accepts requests for changes to the service and drives continuous improvement.

The lifecycle of the service delivery lifecycle value chain

The service delivery value chain is analogous to the supply chain inside most business models. This set of activities is sometimes mistakenly called "IT operations", but in reality it is much more than that. While the activities associated with managing the operational production and execution of services indeed are performed in this chain, it is also serves as the fulfillment arm for all individualized service offerings. The service delivery chain consists of these phases:

Plan: Planning starts with a forecast from the BRM chain that can be translated into capacity, availability, and performance requirements. Also considered are the human resources necessary to manage new projects coming in from the service design chain. In the event of an acquisition in the business, the service delivery chain must be able to plan for a smooth integration of systems and people into the enterprise.

Source: Planned capacity requirements are addressed by products and/or staff. The appropriate resources are sourced from service design, external suppliers (carriers, ISVs, hardware, etc.) as needed. It is quite common for Service Delivery chain resources to work with corporate procurement to complete activities in this phase.

Build/configure: Individualized service offerings are assembled. This phase also includes activities closely aligned with the service design chain as services are transitioned into the production environment, ensuring proper instrumentation and the operational ability to properly manage the services.

Deliver: Activities associated with fulfillment of the service to the economic buyer are completed. Production acceptance testing, operational readiness and collaboration with service design and BRM chains are also handled in this phase.

Manage: Incident, problem and change requests are handled in this phase. Most of the work is associated with data centers, operations centers and help desks.

The Operating Model That Powers the Value Chains

The operating model describes the core and supporting functions of IT, along with the inputs and relationships with the internal business community, suppliers and partners. The IT operating model also defines the work breakdown structure within each value chain.

The IT operating model has four domains: process, information, automation and resources. Each lifecycle phase of each value chain touches these domains.

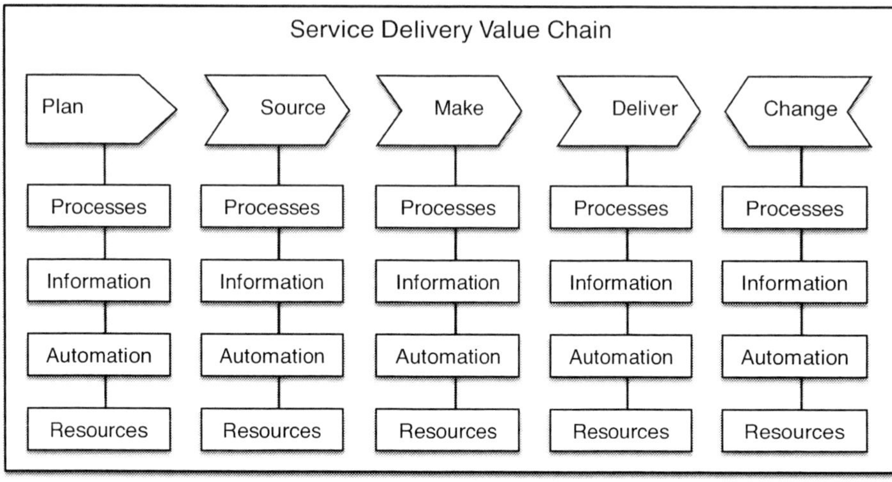

Figure 12 - Each value chain lifecycle is supported by the 4 operating model domains

- **Processes**: Processes represent the flow of work for each phase of each lifecycle. Most processes today are based on best practice frameworks such as ITIL or COBIT.
- **Information**: The data produced and consumed by processes in the lifecycle should be structured around information models that product a "single version of truth" to be used for each lifecycle phase and/or process
- **Automation**: The combination of process and information are used to define, anchor and legitimize requirements for whatever automation is implemented.
- **Resources**: All resources required or involved in each value chain must be defined. Resources include facilities, technology, people and skill sets.

The process domain

The process domain includes the models for work instructions associated with each lifecycle phase of each value chain. Processes typically are modeled, created and implemented using best practice frameworks, such as ITIL or COBIT. Keep in mind that a given lifecycle might leverage more than one best practice framework or methodology. For example, the service design chain may leverage ITIL for the bulk of the work and PRINCE2 for program management activities. Interdependencies among processes are defined and mapped via an information model, which specifies requirements for their integration.

While these frameworks are very effective tools for designing and constructing processes, they do present a problem for the capability-centric IT organization. Since the organization is made up of tribes that don't communicate well with each other, each tribe tends to interpret and apply best practices in their own way. These interpretations tend to harden over time, and can be difficult to change. This can be resolved by moving to a service-centric approach in which value chain-based processes are set in the context of lifecycles, which in turn serve as the governing model for all roles and activities.

For example, in the manage phase of the service delivery lifecycle, there should be clearly defined processes for incident, problem and, change management. These processes can be expressed in flow diagrams and standard operating procedures guides, and are owned and governed by the service delivery lifecycle. Thus, the only people who can modify processes are those involved directly in the value chain.

The information domain

The information domain covers all data produced and consumed by the processes in each lifecycle. This data should be structured around information models that product a "single version of truth" to be used for each lifecycle phase and process. Information pertinent to all phases of a lifecycle is commonly referred to as master data. The most common master data for IT lifecycles includes assets (e.g., software, hardware, facilities), programs and projects, IT services, customers, employees (including roles and skills), suppliers, and partners.

Once again, the siloed structure of the capability-centric IT shop complicates this issue. Member of each tribe define data elements to mean whatever they need to mean inside of their silo, making a "single version of truth" hard to come by. By contrast, in a service-centric model, value chains dictate a common taxonomy and structure for data and thus for information.

Staying with the manage phase of the service delivery lifecycle as an example, the data model should include both master and reference data structures. Data flow diagrams should be linked with process models to offer a complete view of how the mechanics come together in the completion of work.

The automation domain

The automation domain is focused on the deployment of software to actuate processes and minimize the dependency on human intervention. It's important to note that in the context of the IT operating model, the term "automation" refers only to IT processes, and not processes within the lines of business. The combination of process and information is used to define, anchor and legitimize requirements for whatever automation is implemented.

There are three main classes of automation: task automation, process automation and service automation. Each class is more complex than the last, so as you move up from task to process to service automation it becomes increasingly important to have clearly-defined process and information models.

In task automation, tools are used to complete a discrete piece of work, such as adding a table to a database. Process automation involves using both tools and systems to automate a workflow, such as problem management. Service automation involves abstracting work to a service level, perhaps supported by, but not limited to, a SOA within the IT organization. An example might be a provisioning service, which automates any number of tasks and processes towards a discrete service outcome, called provisioning. In the context of the inter-workings of the IT shop, the provisioning service is invoked for many deliverables, such a server, a new employee set up, or an instantiation of a network circuit. The point is, provisioning is defined and automated as a service for IT.

In most capability-centric shops, different tribes have installed their own automated tools, creating new silos and information islands. Using the value chain model as the governing structure for automation helps create an ERP-like ecosystem for the IT organization.

Based on the process and information models defined above, automation for the service delivery lifecycle is much easier to implement. The time-to-value in deployment will improve, along with time-to-productivity of the resources, because the proper context for the automation is in place and everyone knows where they fit in. In big IT shops where there are many moving parts, this is crucial because without understanding the exact context of the operating model, people get lost and try to force automation to do too much, which results in complex, hard-to-support tooling.

The resources domain

The resources domain includes the infrastructure used by IT to run itself, along with facilities such as data centers, call centers and operations centers (such as NOCs). Also included are IT's human resources, its people, their roles and the specific skill sets required to fulfill the lifecycle. The other three domains (process, information and automation) cannot function without these various resources. Each value chain has different requirements in each of these areas, so defining them as part of the architecture helps complete the picture of how each lifecycle phase is carried out.

As an example, resources involved in the service delivery chain would include servers, storage and networks associated with the automation platforms. Roles would include production engineers, operations support staff, and escalation managers, among others. Also specified would be physical plants such as data centers, help desks and operations centers.

Sub-domains provide the operating model with the raw ingredients

Each of these four domains has multiple sub-domains that contain the base level details necessary for the domain to function properly. While domains provide the IT operating model with context for each lifecycle phase, the sub-domains represent the specific details associated with how IT works as an organization. They are the raw ingredients without which services cannot manifest through the various lifecycle phases.

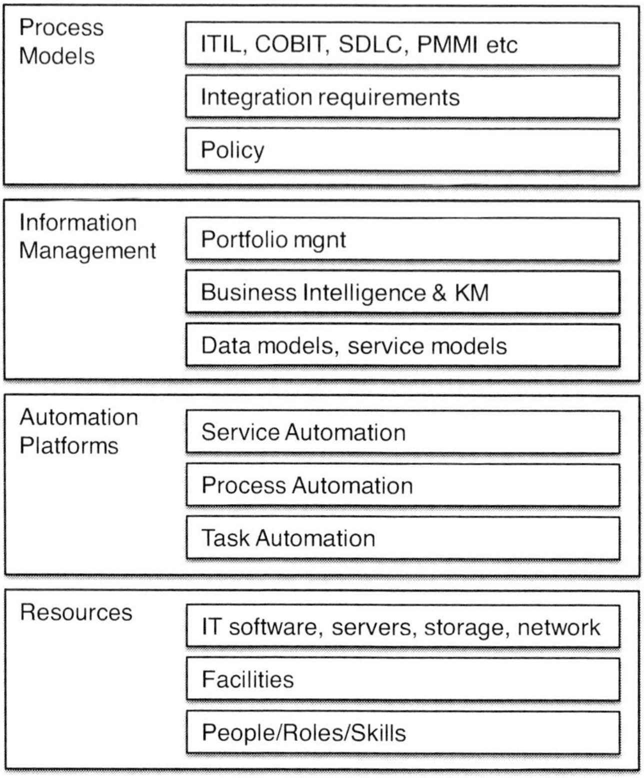

Figure 13 - The operating model has four major domains and multiple sub-domains

For the Process domain, this takes the form of a set of best practice methodologies (ITIL/ITSM, COBIT, SDLC, PRINCE2), along with all necessary integration requirements and pertinent policies. For the information domain, sub-domains include portfolio management, business intelligence and knowledge management. For automation, sub-domains include specific software solutions related to task automation, process automation and service automation. For the resources domain, IT software, servers, storage, networks, facilities, and people are the relevant sub-domains.

The four domains themselves are static—they apply in every case to every value chain. The sub-domains vary by the structure and maturity of the IT function, and enable IT to design and deliver whatever universe of service products are necessary to satisfy its business constituencies across the enterprise. The range of variability possible at the sub-domain level can vary based on organizational maturity, alignment and strategy. In other words, the more effective the IT organization, the more capable it is of defining, implementing and using the sub-domain tools to do its job.

The IT operating model addresses a key weakness in the capability-centric IT organization: the inability to design and manage using a model that will scale to meet the changing needs of the business. Often, the implementation of best practice frameworks such as ITIL fall short because of issues encountered at the sub-domain layer. The IT operating model solves this problem by providing an accurate and dependable frame of reference when set in the context of each lifecycle.

The IT operating model represents the elements of management required throughout the IT function. These elements vary depending on the organizational structure as well as the enterprise operating model. When all the elements are in sync, IT is well aligned with the business to deliver better outcomes and is well on the way to service-centric transformation.

Creating a Data Repository

As we've seen, the EA for IT has four value chains that interact with an operating model made up of four domains and a group of sub-domains. Using this model, a data store can be created that contains all relevant information associated with IT execution. Information placed in this data store is captured as part of the service-centric transformation process, and maintained via standard IT governance to ensure architectural compliance and integrity. This graphic shows a simple reference model for how this repository can be modeled. The EA for IT repository is the place where all content would be stored and managed. Each domain and sub-domain in the IT operating model intersects with each of the four value chains. At each of these intersections, an artifact is created.

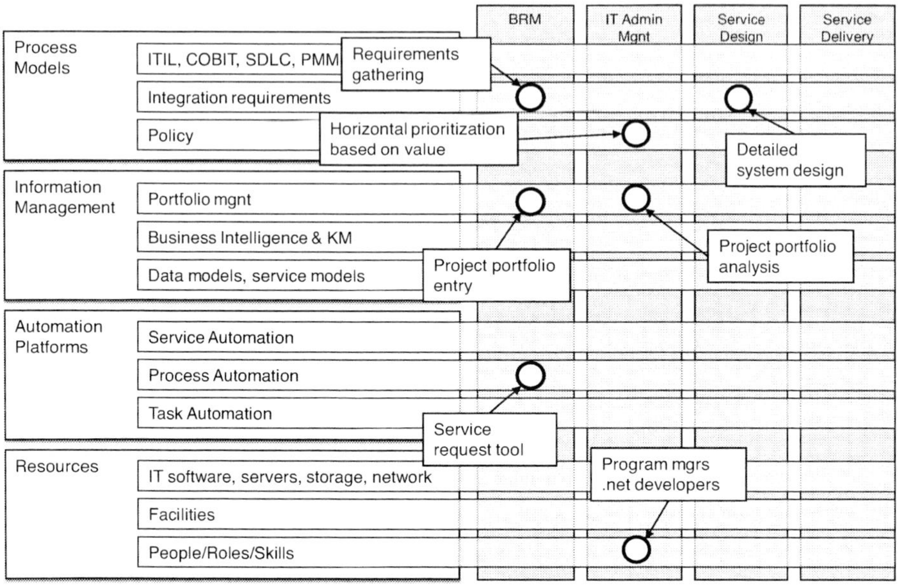

Figure 14 - Where the operating model sub-domains intersect with each value chain, an artifact is created. Articulating artifacts in this way helps the IT organization understand and prioritize work associated with the services, as depicted in the IT strategy.

For example, at the intersection of the ITIL sub-domain and IT administrative management, an artifact captured would be the project prioritization process flow diagram. Where data models sub-domain intersects with IT BRM, an artifact would be the project portfolio data model. Where process automation intersects service delivery, an artifact might be the specification of the problem management system. Using the EA for IT model, every artifact relevant to IT execution can be represented at the intersection of the IT operating model and one of the IT value chains. Any new requirement for service, automation or process changes can be evaluated based on cross-functional requirements rather than tribal demands, and chosen based on how well it delivers a desired and well-defined result.

With the EA for IT structure and supporting repository in place, it becomes easier to promote consistency and to drive visible simplification into the IT model as the transformation initiative is undertaken.

EA for IT Sets the Stage for Value-Based Services

We began this chapter by saying that in order to achieve service-centric transformation, IT's services must be linked directly to the value they deliver.

The EA for IT makes this possible. Everything is defined and standardized—from the smallest units of work to the broadest management and financial measurement processes. Using the EA for IT, an enterprise can at last leave behind the old capability-centric model and undertake a meaningful transformation to a service-centric model. This model is fully aligned with overall business objectives, delivers high-quality consistent outcomes to end users, is fully-measurable across the four vectors of cost, quality, risk and agility, and produces high-value results for executive management.

In other words, the EA for IT gives IT the power to take its rightful place as a key enabler of business innovation.

Pillar Three: Automation and Information

ERP for IT

The third pillar of service-centric transformation is about creating and managing the automation solutions and systems architecture for IT. In simple terms this is the "IT for IT". For the IT organization undergoing service-centric transformation, the goal is to create an integrated suite of applications that enable each value chain to maximize efficiency and effectiveness of the entire IT function.

Today, IT automation generally happens in an opportunistic manner in response to new technologies, techniques, processes or organizational adaptations. Software is selected, implemented and managed vertically within each tribe, which results in a lot of redundant capability across the organization. It's common in large corporate IT shops to find hundreds of different tools in use. These tools run the gamut from device configuration to complex workflow management systems.

Adding to this problem are the many interfaces required, some performed manually, to exchange data both within IT and among IT and the business. For example, IT must interface with the financial systems used to manage budgets and pay bills, the purchasing systems used for procurement of goods and services from external sources, HR systems for managing the IT workforce and contract management systems to govern supplier engagements. These multiple interfaces tend to be unwieldy and very resource intensive.

Together, tool sprawl and complex integration mapping wreak havoc and create isolated information islands. When it comes time for IT to report on events, service level performance or the general IT state of affairs, superhuman efforts are required to harvest data from this wide array of sources, and then assimilate it into some form of useful information. Needless to say, the quality of this data can't be measured on any Sigma scale, and the manual handoffs between automated workflows result in high error rates.

As part of the process of service-centric transformation, automation is created and governed in the context of the enterprise architecture for IT to build efficiency into the design and delivery of each service, along with the management of IT business functions. Over time and through architecture governance, an integrated system and information portfolio emerges that functions much like an Enterprise Resource Planning (ERP) system used in other lines of business.

Applying context and lessons learned from ERP

ERP solutions emerged as a common practice in the 1990's, as part of business process re-engineering efforts within many corporations. Solutions from software providers like SAP and Oracle were implemented to automate processes for most major business functions. These efforts were catalysts for other significant IT-led business improvements in areas such as business intelligence (data warehousing) and business process management (enterprise application integration), as focus shifted from simply automating processes to information management and the integration of extended value-networks (e.g., suppliers, manufacturers, customers). Because of advances in technology and organizational maturity gains, it is common to find corporations who are now on their second- or third- generation of ERP implementations.

At the highest level, ERP systems focus on structuring and automating most aspects of an organization's work, such as tasks, processes and information exchanges. Requirements for modifications and customizations are grounded in an enterprise architecture, which includes corporate information models. This approach produces an authoritative source model for systems and information, which when implemented, streamlines the performance of tasks, eliminates redundancy and moves data quality levels towards the Six Sigma range.

The design and implementation of an ERP system is generally framed by the enterprise architecture for the business—a top-down approach. In some instances, the act of implementing an ERP system actually serves as a catalyst for changing the behavioral, cultural and organizational attributes of a company in a more bottom-up fashion. In either case, implementations are, by nature, disruptive to the status-quo of the business itself, which has given it a negative image in some quarters.

Rather than consider the positives and negatives of ERP systems, our objective is to offer a meaningful description of how these systems are underpinned by a business architecture, and thus how ERP is useful paradigm for approaching the transformation of the IT organization. Clearly, IT can leverage the business-like rigor and structure applied to ERP systems for its own automation. Effective ERP for IT implicates the following characteristics:

- It is set in the context of the structure and operating model of the IT function: the EA for IT
- It draws heavily from the automation and information domains in the operating model
    - Uses common data architecture and models
    - Conforms to an authoritative source model: one version of asset, service cost, etc.
    - Consists of standard platforms and infrastructure: the IT infrastructure for IT
    - Drives towards model-based design and development methodologies: SOA for IT, reusable components
- It comprises a portfolio of systems and information rather than a collection of tools
    - Uses information to facilitate IT and business decisions
    - Invests in areas providing the highest returns
    - Maximizes leverage across value chains
    - Uses buy vs. build strategy: SaaS, off-the-shelf packages, etc.

To explore this further, we will drill down into the automation platform and information management domains depicted in the operating model.

IT for IT: Distinguishing Layers of IT Automation

The automation of IT should be designed and implemented in relation to each value chain in the EA for IT. Remember that for each value chain, there are three major categories of automation: task, process and service. Each category has its own classification of complexity and breadth of coverage. Progression through each level requires greater understanding of the IT structure, and is more difficult and time consuming to implement.

		ERP for business examples	ERP for IT examples
Service	Service automation provides mechanism for the IT consumed services	Order status Shipping service Quote/configure Calculate sales tax	Order service Provisioning Service status Project status
Process	Process automation controls business workflow execution often using best practice models	Pick/pack/ship Factory floor controls Demand planning Order management New employee onboard	Problem management Project management Service design New employee onboard Release to production
Task	Task automation focuses on discreet, often repetitive actions, reducing complexity while leaving the right amount of control	Configure order Inventory lookup Post G/L transaction Calculate sales tax Label printing	Service order status Upgrade Windows Change router configuration Add user password Extend table space in DB

Figure 15 - Task automation, process automation and service automation support ERP for business as well as ERP for IT

Task Automation

Task automation is focused on automating work performed by individuals within each value chain. Task automation can be embedded inside a technology solution such as a server, delivered as a software package or through a subscription-based service. In addition to automating tasks associated with services, this category also focuses on data produced and consumed in the various stages of the service lifecycle.

Task automation is the level where the majority of IT tools exist today, and the bulk of the information is produced. Tools designed for task automation are generally integrated with other tools within the ERP for IT ecosystem through some form of higher level IT management system. Figure 16 reflects examples of task automation for each value chain.

Task Automation

IT Administrative Management	Business Relationship Management
▪ Financial forecast ▪ Submit RFP ▪ Pay Supplier	▪ Order status ▪ Report SLA ▪ Change request ▪ Order new desktop
Service Design	**Service Delivery**
▪ Add source code to library ▪ Configure server ▪ Install blade server	▪ Change router configuration ▪ Monitor server ▪ Dispatch technician ▪ Change password

Figure 16 - Task automation examples for each of the IT value chains

Process Automation

Process automation systems control workflow associated with the different stages of the value chain lifecycles. While task automation generally relates to individuals and their roles, process automation crosses roles and can also cross value chains. Process automation solutions are critical for improving speed, efficiency and predictability, but obviously they're only as good as the process models behind them. Therefore, many process automation tools today are being designed using best-practice frameworks such as ITIL, COBIT, PRINCE2, and CMMI.

Successful process automation requires shifting behaviors across tribes and silos to create a uniform system for managing each IT service over the course of its lifecycle. Process automation often requires integration with task automation and other process automation solutions. Most process automation suppliers provide out-of-the-box methods for achieving this integration. However, as the scale and complexity of the IT ecosystem has grown, these integration methods have become more difficult to implement and manage. Responding to this, software suppliers in this area are moving towards more mainstream integration practices such as EAI and SOA.

Process Automation

IT Administrative Management	Business Relationship Management
▪ Employee performance review ▪ Change advisory board ▪ Build plan of record ▪ Standards compliance review	▪ Requirements management ▪ Project management ▪ Data modeling
Service Design	**Service Delivery**
▪ Application development ▪ Project management ▪ Network design ▪ Database design ▪ Server build	▪ Problem management ▪ Incident management ▪ Change management ▪ Escalation management

Figure 17 - Process automation examples for each of the IT value chains

Another component of effective process automation is an authoritative source model, similar to those used in business ERP systems. An authoritative source model is based on the idea that there must be one system of record for controlling a named dataset. This fosters a reliance on organization structure and alignment in order to achieve the desired end state.

For example, if the structure and operating model for IT is decentralized across different lines of business, each having its own problem management solution, multiple sources of problem data are created. When a problem crosses IT tribal boundaries, that data must be converted in some way to pass from one system to another, or be handed off manually, in order for the problem to be solved. An authoritative source model addresses this issue, but requires a level of compromise and governance that is not typically found in IT organizations. An authoritative source model also helps the IT organization move toward a common information architecture.

Service Automation

As mentioned earlier, this class of automation involves abstracting IT work to a service level, so that the IT mechanics are embodied as a set of discrete, repeatable, efficient services designed to automate the business of IT. An example might be a provisioning service, which automates any number of tasks and processes towards a discrete service outcome, called "provisioning". This provisioning service can be actuated against any number of IT-provided services, ranging from building out a server farm, to on-boarding a new employee, or configuring a network circuit. In each of these examples, a common provisioning service is utilized which automates and accelerates the instantiation of the deliverable.

To better understand this concept, let's first look at how service automation is typically applied to services consumed by lines of business. Mainstream application services used by the business are designed with re-usable components set in the context of goods or services being sold and delivered to end-customers. Techniques such as SOA and Web services frequently are used as a means accomplishing this goal. Using a simple example, instead of building a shipping application for the processes and tasks around each different product or product line, a generic shipping service can be created in such a way that allows users to ship any product to any end-customer. This shipping service is accessible via any authorized application, and from a multitude of devices, such as a PC, cell phone, or PDA.

Without the shipping service, any application involved in the order fulfillment process must understand and codify relationships with customers, logistics partners and manufacturing—which adds complexity, latency and cost to the logistics process. The generic shipping service introduces a level of abstraction that manages these relationships on behalf of any order fulfillment system for any product family.

In the same way, service automation extracts the essential elements of a value chain, models them around the service, and offers them in a consistent way, thus creating new and innovative ways of connecting the business with its suppliers and customers.

IT has not applied this practice to the design of the services it consumes internally. In fact, IT tools such as management software have historically not lent themselves to this model, because most of them have been designed to execute at the task and process levels.

From here, we'll focus our service automation discussion only on the services IT needs to perform its work inside each value chain, previously referred to as IT-consumed services.

When the service automation concept is applied to the IT automation environment, services can be identified, modeled and designed to connect with other IT entities, IT suppliers and consumers of IT services. Service automation enhances predictability and efficiency for offering, delivering and managing services. For example, when it comes to adding capacity to a new or existing environment, automation can vary by technology domain. If server capacity is needed, server provisioning tools are used by system administrators to complete tasks and feed data into an appropriate request management workflow engine. If network or storage capacity is needed, then automation solutions for those technology domains are used to complete the tasks. Most of these tools aren't integrated with any workflow controller so they require manual intervention to connect a work request (aka a ticket) with the right processes, tasks and humans.

Using service automation, the activities associated with adding capacity can be codified into a generic provisioning service accessible from a variety of IT systems. This means that computer, network and/or storage capacity can be added using the same provisioning service. In addition, this provisioning service can be accessed via any approved application to integrate with various workflow control systems and task automation, thus eliminating dependency on human intervention to connect the tasks and processes. It is not a stretch therefore to consider other service automation opportunities in IT.

Figure 18 shows some examples of IT services that can be developed (or purchased) to deliver the maximum return.

Service Automation

IT Administrative Management	Business Relationship Management
▪ Approval service ▪ Billing service ▪ Payment service ▪ Employee skills lookup	▪ Service catalog service ▪ Service quote ▪ Configure order ▪ Service status
Service Design	**Service Delivery**
▪ Estimating service ▪ Service creation ▪ Project status	▪ Order service ▪ Change service ▪ Service restoration ▪ Authorization / authentication

Figure 18 - Service automation examples for each of the IT value chains

Think about the hundreds of calls that the typical help desk receives daily about poor e-mail performance, or not being able to access a certain shared file. If there were a service status service, each employee could examine the state of the service themselves without having to engage an IT resource. If a project manager could use a skills lookup service to check resource availability for her project based on skill set needs, she could avoid countless e-mails, meetings and phone calls searching for the appropriate members to staff the project team. A project status service could reduce the amount of time team members spend on status reports, enabling them to spend more time actually doing project-related work. These are just a few simple examples of how service automation can benefit the IT shop.

Let's briefly review our discussion of task, process and service automation. Task automation generally is implemented inside of a tribal silo to serve a technology domain or IT function. Process automation also is implemented vertically, but with some integrations required to move data from one silo to the next. Service automation is implemented horizontally because it focuses on the service itself and progression of that service through a pre-defined lifecycle. This lifecycle spawns processes and tasks within the silos, so while the work doesn't change, the context in which the work takes place is radically different.

Information About IT, for IT

The next operating model domain explored in the ERP for IT is information management.

Information produced by IT is nothing less than the DNA of a business. Every customer transaction, supply request, facility, and product design is captured and can be seen through the lens of the IT ecosystem. However, most companies don't treat IT information this way. Instead, this data stays locked behind the walls of IT, stored inside of niche tools, accessible only by specialists who work hard to ensure that it is used almost exclusively by IT people. In most companies, the closest IT data gets to the business is when it is assimilated into reports that measure IT's performance against its stated service levels goals.

You don't have to be a visionary to recognize that this IT information would be very useful to business leaders in making policy, organizational, financial and operational decisions. For example,

- A manager responsible for manufacturing lines with 1,000 employees is under a lot of pressure to reduce the cost of his operation, and he thinks his only option is to eliminate jobs to meet his financial targets. Imagine if he were able to see that his line of business was spending $10,000 per month for cell phone usage and $25,000 per month for e-mail service. Wouldn't this cause him to make some behavior changes and policy decisions that might save some of his workforce?

- What if executive committee members could see that the business was incurring a one-dollar-per-product overhead charge on a ten dollar product cost to cover the utility bills associated with their five data centers? Wouldn't they be motivated to eliminate some of those facilities?

- Suppose a CIO could show that, because of business process complexity, it costs five dollars in systems transaction charges to process a single order from quote-to-cash. Wouldn't that create movement toward simplifying the business process?

These are simplistic examples, but they make an important point: data that could help expose these conditions and drive better business decisions is trapped in multiple, disparate parts of the IT systems. Obtaining this data is difficult enough; turning it into meaningful business information is virtually impossible because the "information mindset" that exists in every other line of business does not exist within IT.

Service-centric IT transformation is a catalyst for creating an information architecture that supports the ERP for IT. Let's briefly examine the sub-domains of information management.

The information hierarchy

The first sub-domain of the information architecture is the information hierarchy, which has two levels: master data and reference data.

Master data is a clearly defined set of authoritative source information that is maintained in one place is consumed by many other systems within the value chains. Data models for master data must be closely governed and repositories should have the highest levels of data integrity (approaching Six Sigma). Authoritative source systems for the production and management of master data must be named. There are typically five to ten master data sources in an ERP for IT ecosystem. Examples of master data sources include assets, services, people and projects.

Reference data is derived from action taken on master data through processes or tasks. Data models for reference data generally are driven by commercial software suppliers. So it's important that there be a data management architecture that rationalizes the data elements into models at different levels. Examples of reference data include configuration information, project status and service requests.

Information lifecycle management

Much like a service, information flows through a lifecycle. This lifecycle can be characterized by the various containers described below. At each stage, data and information has value to different stakeholders. The IT organization's ability to capture, store, and make intelligent use of this information is a make-or-break proposition: the information is the mirror by which IT can judge its efficacy as a service-centric IT shop.

Online transaction processing (OLTP) data is closely linked with a transactional system. Problems, events, incidents, change requests are all good examples of OLTP data. This data is not typically aggregated or summarized. Producers and consumers of this data are the IT staff performing their daily tasks.

Operational data stores (ODS) are lightly summarized and transient data, typically covering a defined time period, such as a day, week or month. The configuration management database (CMDB) is a good example of an operational data store. Overall performance of the SAP order management system (network, server, storage, application domains) over the last week would be a good example of ODS data. This data is generally produced by the OLTP systems and consumed by planners, supervisors and administrators.

Data warehouses collect highly summarized and highly aggregated data used for trending and analytics. The data warehouse for IT can be the same one used by the rest of the business. Transaction analysis of the shipping ecosystem, including EDI, logistics applications, and infrastructure, for the year would be a good example of how DW data is used. This data is typically produced by both the OLTP systems and the ODS and is consumed by planners, IT managers and the business as a whole.

Data marts are not physical entities; rather, they are simply different views into a data warehouse used for specific purposes by IT or a line of business. For example, if the business wants to monitor performance of its supply chain lifecycle, it could organize a data mart to view the servers, storage, applications and network performance for these business processes.

Data architecture

For our purposes, a data architecture for IT is made up of logical and physical data models along with the database designs and integration architecture. Most IT automation solutions have built-in data models and database designs. So as the automation portfolio has expanded in IT, so has data model disparity. These disparate data models make integration and information exchange cumbersome and complex. Today, companies are attacking integration between solutions in various ways to facilitate end-to-end process automation. Complex APIs, evolving Web services and even enterprise application integration solutions are some of the things being tried in order to accomplish this task.

This has made the CMDB a hot topic in IT management software circles. When it comes to service-centric transformation, the form and shape of the data architecture should begin with the IT services themselves. Modeling of the services should serve as the foundation for the data models and data architecture of the automation ecosystem.

Creating an ERP for IT

Actuating the IT automation and information management domains of the operating model in should be done in the same way as it's done in any other line of business. This usually isn't the case, however—they are most often undertaken opportunistically and inconsistently, which results in a lot of redundant tools, poor data quality and a complex fabric of poorly-integrated systems.

There are many operational efficiencies and economic benefits in building an integrated IT systems architecture that resembles that of the business ERP system. Using the operating model domains as a guide to focus the IT shop on the most critical supporting mechanisms, you can start to harness the right structure with which to undergo the transformation to a service-centric model.

Of course, it helps to have a recipe—a set of directions for building a service-centric IT shop in a way that shows incremental success, and incremental value. We'll cover this in detail in the next section.

Pillar Four: The Transformation Recipe

The service-centric IT shop is built around a line of IT services, which manifest through a series of provider and consumer relationships, or economic exchanges. Not unlike any other service-centric line of business, IT has to position and structure itself to adapt to this type of ecosystem.

But how do you get there? Is it really possible to undergo a shift to align with the service-centric economy, and to expect that you'll achieve lasting results?

The short answer is: yes.

It happens by bringing about changes in thinking and behaviors, and by creating the core structures within IT that naturally generate the right interdependencies and synergies between people, structure and technology. This is the essence of service-centric IT, and the vehicle for doing so is the IT transformation recipe—a quick, iterative, yet rigorous process that yields short-term progress toward a strategic and lasting end state, which includes:

- A well-defined set of IT services linked with business outcomes, results, and value
- An organization structured around value chains and lifecycles
- An operating model designed to fully leverage process, automation, information and resources

The service-centric IT model offers a common set of IT services that are linked to business outcomes. It is financially transparent and has a reality-based cost structure, so the business can follow the money it spends on IT, and measure the value it receives. IT can reduce spending on maintenance and recalibrate its efforts to provide more of the innovations the business depends on for growth and differentiation in the marketplace.

The IT organization can permanently position itself as both a tactical value-based provider of choice for every internal department, and a strategic leader in driving overall business performance.

Most importantly, the service-centric IT organization can permanently position itself as both a tactical value-based provider of choice for every internal department and a strategic leader in driving overall business performance.

Steps to a Transformed IT End State

The IT transformation recipe is a set of repeatable steps that provide an achievable path toward transforming IT from a capability-centric state to a service-centric end state.

There are many formal methodologies which define best practices for handling different kinds of initiatives: project management methodologies, system design methodologies and even service management methodologies. The IT transformation recipe is both simpler and more holistic. The recipe is based on real-world IT experience, and each step in the recipe helps bring life to a new, scalable structure that includes the elements discussed in previous chapters, including a defined services portfolio, value chains, operating model, and systems architecture for IT.

There are five steps in the recipe which, when executed sequentially with a few services (or even just a single service) as the variable, begins to actualize the service-centric end state.

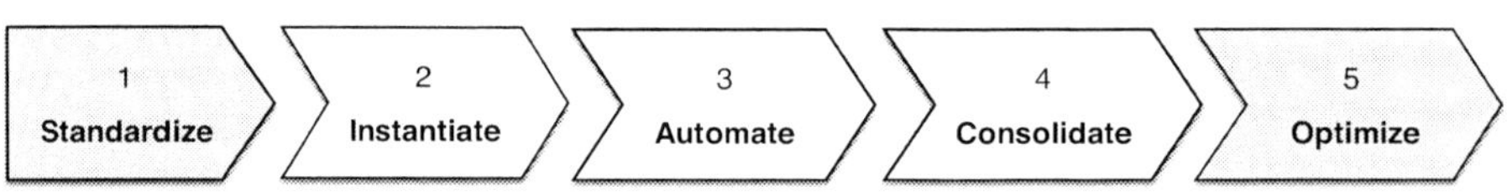

Figure 19 - the steps of the transformation recipe

The five steps are:

- **Standardize**: planning, modeling and homogenizing attributes of the service, the value chains, and the operating model
- **Instantiate**: creating an instance of the service, and actuating on elements of the value chains and operating model

- **Automate**: using software to mechanize the processes surrounding the service, value chains, and operating model
- **Consolidate**: migrating the service and artifacts of value chains and operating models from existing locations and form factors into the transformed environment
- **Optimize**: tuning the service in the context of the value chains and operating model, and recalibrating the remaining resources

Here's how it works: by going through the steps using a few services in the first of what will be several rounds, the enterprise architecture will begin to emerge as a by-product of the results.

	1 **Standardize**	2 **Instantiate**	3 **Automate**	4 **Consolidate**	5 **Optimize**
Description	Define an authoritative model for the service, as well as associated EA for IT attributes	Architect, engineer and install the service environment	Implement technology to automate the value chains, operating model and service lifecycle	Migrate existing services to the new transformed state	Harvest assets and resources for reuse; Recalibrate, tune and refine existing and transformed landscape
Value	Generates standardized models for reducing cost and complexity	Establishes the structured end state environment for service-centric IT	Streamlines work associated with tasks, processes and services and creates a single version of truth for IT information	Minimize service footprint and reduce cost of management and delivery	Exposes the economic and qualitative benefits of the transformation effort

Figure 20

In order to get a feel for how the IT transformation recipe works in the real world, let's choose an example of a common IT capability and transform it into a service using the recipe as a guide to a transformed end state. Here's our example situation:

The capabilities that we will address are the servers, storage, network, applications, people, processes, automation information and facilities that are used to deliver various types of web-hosted business services. In this example, ACME Corporation has these capabilities spread across three worldwide regions, and managed by a wide range of people.

We will create a single, standardized web hosting service using the BEA WebLogic platform to handle the transaction workloads. The service will be positioned in the infrastructure category of the service catalog. It will be centrally located in a data center in Omaha, Nebraska.

It will be designed as a configure-to-order service, accommodating e-stores and knowledge sites accessible from internal and external users around the world. It will have a basic set of features with a limited number of configurable options that relate to security, storage and availability. Because it's a configure-to-order service with some pre-determined components, what it doesn't accommodate is changes to types of hardware, software or anything else that doesn't meet the boundary conditions set forth in the features and options listed in the service catalog.

We'll use figure 20 as our guide as we go through the steps. Recall that our web hosting service is considered an IT-provided service, and thus is consumed by a line of business. We will discuss what happens as you take it through each of the five steps. In each step, we'll consider the service itself, the value chains that affect the service through its lifecycle, and the operating model domains and sub-domains that enable the service lifecycle.

It's important to note that this is a build-as-you-go approach. In other words, rather than building out all of the value chains, operating model, automation and information beforehand, by having the structure and intent of the end state model in mind, it will actually emerge as a result of going through the five step process.

A cautionary note: in the ultra-complex world of the modern IT organization, the recipe might seem at first glance to be ridiculously oversimplified. But it is this simplicity that enables the IT transformation recipe to succeed where all those previous initiatives have failed.

That same simplicity should be used to ensure success in applying the IT transformation recipe. A lightweight approach is best to avoid atrophy or implosion while maximizing speed and short-term value. Specifically, the optimal strategy is to work iteratively through the recipe with the service(s), instead of trying to achieve wholesale transformation of technologies, facilities, people or processes. By leading with IT services, these larger IT vectors will be pulled along as activities progress. Choose services that are simple or that the IT organization is well-experience with already—perhaps something the organization is already close to packaging as a service. This will produce early successes and set the tone for the iterations to follow.

As each subsequent service passes through the recipe, IT can leverage and re-use artifacts, structure and deliverables from prior passes, increasing speed, building consistency and simplifying ongoing support, creating a transformation snowball with its own momentum.

Being 100 percent complete in transforming the first few services is not necessary. Aim for 60-70 percent accuracy in the initial passes. Time, experience and maturity improvements with the service-centric model will enable continuous improvements to be made as needed. Furthermore, since the recipe mechanics are iterative, improvements to structure, automation, processes and roles will evolve with progressive passes. What's important is to start with three things: a foundational model for the EA for IT, a basic service taxonomy and a financial target.

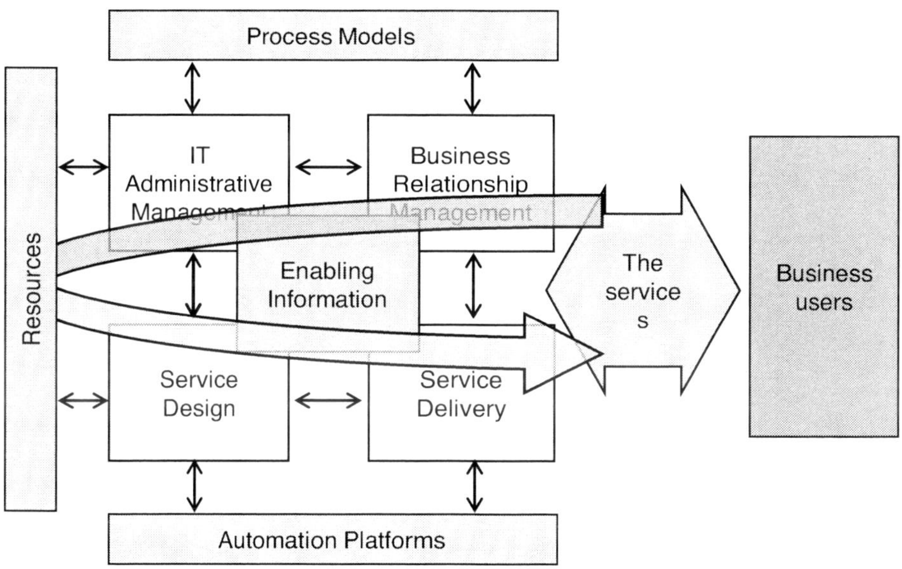

Figure 21 - Each service travels through the value chains and operating model for each of the five steps of the recipe.

By going through every service in a step-by-step repeatable way, IT is able to create a new and more business-focused service portfolio while also describing and defining the value chains and associated operating model components.

Step 1: Standardize

The first step is to standardize. The intent of this step is to take a service or small set of services through a lifecycle that drives compliance with and alignment around a common set of components, policies, best practices and other EA for IT elements, as well as compliance with the processes and systems required to design, provision and support the service.

The main activity invoked in this standardization phase is planning and modeling. That is, each element of the end state is considered with respect to the web hosting service, and once completed, we'll move into step two where it is instantiated.

Standardization delivers an authoritative model or blueprint for each service. Standardization is key to achieving cost reduction and efficiency because it eliminates the number of disparate elements involved in creating and delivering the service. This reduces the number of people and amount of investment required to support and maintain the service over its lifecycle.

Standardizing the service

The first thing we'll standardize is the service itself. Standardization introduces boundary conditions with which the enterprise will manage its IT business. While seemingly limiting flexibility, it introduces rigor that has a lasting, positive affect on cost structure, speed, and predictability of business outcomes. Standardization enables the service-centric IT organization to accomplish something that capability-centric organizations only do rarely: develop a clear and accurate value proposition for individual services. This not only helps IT understand and articulate the value of its services, but also serves as the foundation for cost savings and recalibration of the maintenance to innovation spending ratio.

Implementing a standard service taxonomy enhances communication between business and IT, so problems and issues can be identified and resolved faster. Standardization creates an accurate and up-to-date asset inventory, and reduces the number of images and platforms IT must support.

Finally, with each standardized service, IT gains the knowledge, experience and reusable elements it needs to keep the transformation process going.

The first step in creating any standardized service is the development of a model that fully describes its taxonomy and specifications, for example:

What is the service name?	BEA WebLogic-based web hosting service
What service category would this fall under?	The hosting service is considered an infrastructure service since it is delivered as a platform to be consumed by other parts of the IT organization
What are the features of the service?	Example features include the amount of processing power and storage allocated to transactions, security policies, the ability to access internally and externally
Who are the economic buyers?	Line of business IT teams, application developers
What is that delivery model?	Centralized or decentralized support organizations, help desk, service catalog and pricing information
What service level options are available?	Gold, silver and bronze for availability and response time

The various levels of the services' service level agreements will have an associated cost implication, technology implication and operational process implication. Modeling these service level descriptions helps the IT organization better understand how to prioritize workloads and helps the business better understand what to expect for its investment. For our web hosting example, we will offer gold, silver, and bronze service levels. Each of these levels has different attributes around accessibility, availability and responsiveness from support. These service levels are part of the configure-to-order composition.

Next, the technology related elements of the service must be modeled. In our hosting example, the base product would consist of BEA WebLogic software, along their compute, storage and network requirements.

From there, the task becomes to collect and document the standardized set of other configurable options that will be made available to buyers of the service. Such options might include direct access by suppliers, partners, or perhaps PDA access. It's important to fully consider the extent of, and implications for configuration options up front in this step.

Next, the facilities necessary to deliver the service are defined. Does the hosting service need to live in a certain building? Should it be physically located close to the manufacturing line, or in a specific geographic location? In this example, we have decided to host in a single facility, the Omaha data center.

From these activities, a service model emerges, and a pricing model for the service can be developed. Cost and price, for our purposes here, are two different things. Cost is what IT must pay to create, deliver and maintain the service over the course of its lifecycle. Price is what the economic consumers of the service must pay IT to receive the service. The cost/price differential doesn't represent the profit IT makes on a given service. Rather, it covers the expenses associated with delivering and managing the service on an ongoing basis.

Generally, the lifespan of a service is three to five years from inception to the time it undergoes some form of renewal or replacement. Over that period, IT must cover the cost of ongoing administration, design, increased hardware capacity, support staff, license fees and operational expenses generated from expansion and unforeseen issues such as breakdowns or outages. Those costs, which typically range from two to ten percent, are added to the base cost, and the total becomes the price for the service. In addition, the price should vary by service level to cover the expenses associated with offering service at the bronze, silver and gold levels outlined above.

Once the modeling and standardization of the service definition is complete, it's time to move on to standardizing the value chains.

Standardizing the value chains

Recall that at the highest level, the EA for IT is made up of four value chains: IT administrative management, business relationship management, service design and service delivery. Each value chain is expressed as a lifecycle that consists of a series of phases. For any given service, all four of the value chains will be touched at some level, and therefore standardization must be done at every touch point, as well as across the operating model, which we'll get to in a minute.

At this point, because we are focused on standardizing, which invokes modeling and planning, the structure of the phases for each lifecycle in the value chains need to be defined. This is an important step, because it will be the governing structure for all actions that happen to subsequent services as they traverse through their lifecycle. Lines of demarcation will be drawn between different parts and functions of the organization along with interdependencies.

To illustrate how a service flows through each value chain in the context of forecasting demand for the upcoming year, let's look at how the hosting service traverses the value chains. First, roles aligned with the BRM value chain should be engaged with their respective business units to gather demand for all hosting services in the upcoming cycle. The addition of employees to a particular business unit or system-generated transaction flows that consume hosting services would be captured as incremental demand to the existing service.

This information would be passed to the service design chain for assessment of the demand impact on the service, such as increased servers, licenses, storage and network capacity. Service design resources then would work with service delivery chain teams to determine how the demand can best be met. Simple capacity increases are typically handled exclusively by the service delivery chain. More complex increases—those that might require changes to the service underpinnings—would most likely become project work performed inside the service design chain.

With the assessment completed, the rationalized demand along with the estimated costs would be submitted to the BRM chain, which would review them alongside all the other demand signals to arrive at a horizontal prioritization of investments across the service portfolio. Once prioritized, any project work associated with increased hosting demand would be assigned to service design or service delivery in order to be fulfilled.

Standardizing the operating model

As you go through each of the value chains, there are a variety of touch points with all four domains in the operating model: process, automation, information and resources. We'll now look at each one of those domains.

Let's begin with process model standardization. Because we are targeting a standardized service and moving to away from the customized nature of the build-to-order philosophy, the associated processes for controlling how work gets done can be collapsed into simpler, repeatable models.

Generally, process models are best represented by flow diagrams. Most of the necessary process models for the design, delivery and management of any given service already are available as part of best practice frameworks such as ITIL, PRINCE2, COBIT and others, so there is no need to spend a lot of time and effort re-inventing them. Each modeled process interacts with the four Value Chains and operating model of the EA for IT:

- IT administrative management: processes include accounting, budgeting, cost recovery, procurement, demand and supply planning, HR, training, contracts management, among others.
- BRM: processes include demand planning (vertically within the individual business units) requirements management, service level management, service portfolio management and project management, among others.
- Service design: processes include project and program management, sourcing, design, testing, release to production, among others.
- Service delivery: processes include order management, provisioning, production acceptance, change management, problem management, configuration management, performance and capacity management, among others.

The principal objective of creating standardized process models inside the EA for IT is to be able to re-use the models again and again as multiple services are transformed through iterations of the recipe—moving ever forward towards configure-to-order behavior. Proposals to deviate from a standard can be rationalized against the EA for IT content. Without the overarching structure of value chains linked to the EA for IT, different processes are developed by each tribe. For example, the problem management process for the network team often differs from the application development team for no other reason than learned behavior or culture within each team. Using the EA for IT as a guide, standardization at the process level can be tied to each respective value chain, which makes for greater consistency at the process layer.

The next layer of standardization involves building an information model around the service—defining the information produced and consumed in the design, delivery and management of the service.

Think about how this plays out in a restaurant: When talking with the hostess, there is an information exchange around, for example, how many people in your party, need a highchair for the baby, and how many kids menus you need. When talking with the waiter there is yet even more information exchanged: what would you like to eat, how would you like it, what to drink, etc. These two examples reflect information models that exist in each of their respective value chains.

In IT as well, there are information exchanges in each value chain. Left un-modeled, the resulting deliverables can fall short of expectations and the reporting associated with the performance of the service or the organization can be a daunting task. Information modeling for IT is often ignored and typically driven by the collection of tools in the ecosystem. However, in the service-centric transformation, there is a chance to correct the situation and improve the fact basis used for making decisions, and overall measurability of the organization. Standardized information models should exist in each value chain. Let's take a closer look:

- IT administrative management chain: key performance indicators needed to measure the performance of the service include skill set requirements, suppliers, and service costs
- BRM chain: information attributes and entities associated with an order for this service, such as number of users, locations, transaction profiling
- Service design chain: information includes compute, network and storage specifications; hardware profiling
- Service delivery chain: information associated with the fulfillment and ongoing service support, such as events, incidents, problems, user profiles.

The deliverable is a standardized information architecture for the service, which most often takes the form of data flow diagrams and logical data models.

Using our web hosting example, we would have an information model for the service itself consisting of name, specifications, consumers, delivery time and suppliers. We would also have information models for problems, events, orders, and other data entities associated with the service lifecycle.

The third layer of operating model standardization is automation. Keep in mind that in this phase we are modeling the automation instead of implementing it. Modeling requires an examination of how the organization functions in the context of its deliverables and perhaps some pre-existing boundary conditions.

Think about your favorite fast-food restaurant. With the emphasis on speed of delivery and volume sales, its likely that almost every task and process in the place has been automated. Where to place the automation and how to use it didn't happen by accident. Instead, it's more likely that by studying behaviors, processes and tasks in the context of the services offered, automation was modeled before it was put in place. IT is no different in many respects when it comes to the need for automation modeling. Yet, what typically happens is one of the tribes starts talking about requirements for automating a certain set of processes or tasks to make life easier. Some requirements are jotted down and it's off to the races in search of a solution.

In the service-centric IT, there are as many as three parts to creating a standard IT automation model: task, process and service automation all framed in the context of the service as it flows through the different value chains. What follows are some examples of the automation models that should exist within each value chain for a given service:

- IT administrative management chain: project and portfolio management systems model, budgeting, contracts management, HR
- BRM chain: systems model for requirements management, demand, and project management
- Service design chain: software development and quality assurance model, release management
- Service delivery chain: event, incident, problem, change, knowledge, and service request management

Again, this is not about creating the automation itself. It's about identifying and developing a standard model for what is going to be automated and how. The deliverable is a schematic of the automated ecosystem, a general set of requirements, as well as any associated RFPs for software vendors.

In our web hosting example we would need to model how the service would be monitored, all the way from the component level to the user or consumer level. That's pretty easy to see. But when it comes to modeling how orders for the service are going to be captured, and how they will flow through the different value chains to fulfillment, it's a little more difficult. It requires aligning the different process models and information flows and determining where the automation opportunities exist, what's already in place, what needs to change and where the gaps are. What comes out of the activity is the systems architecture diagram for the service over the course of its lifecycle.

The final layer in our standardization step is resources, which deals with modeling the roles, the technology, the facilities and the other enablers of the transformed service. The number of people, their roles and skills, the technology elements, the facilities and so on should all be established in the context of the service being offered. If a restaurant is set up to cook Mexican food, its highly unlikely that they are going to be equipped with the ingredients, people or even the facilities to deliver sushi to customers. The same is true in IT. However, because we've modeled our processes, information and automation around the service, modeling the resources in that same context will ensure that we are able to design, deliver and manage the service in a highly predictable, highly repeatable manner throughout its lifecycle. Here again, standardization of the resource models occurs within each value chain. Here are a few examples:

- IT administrative management chain: Roles in this area include project management professionals, financial analysts, HR analysts, procurement generalists and contracts management staff perhaps in the legal department.
- BRM chain: roles include program managers, business relationship manager or business liaison
- Service design chain: roles include project managers, systems engineers, network engineers, and storage engineers
- Service delivery chain: Roles here include the data center operations teams associated with keeping the hosting servers, storage, network and software up and running.

Next, we need to model the technology resources required for the enablement of our service. These include but are not limited to the following:

- Computing platform for all service components. The intent is to define or conform to a standard computing platform on which to host front-end, back-end and middle-tier computing workloads.
- Storage platforms used, in the case of web hosting, for the retention and retrieval of data and information. These platforms should be able to serve the purposes of as many services as possible.
- Network platforms used by the service at all levels including, local-area, wide-area, storage-area along with various types of connectivity (wire-line, wireless, etc.).
- Standard releases of operating system software
- Standardize the technology resources for the IT management platform—the IT technology platform for IT, on which IT management software runs

In addition to the human and technology resources, the facility requirements of the service must also be defined. These facility requirements are generally a by product of the technology form factors. Power, HVAC, zoning and physical security aspects are all elements of the facility model. These standards rise in importance as more and more services are transformed. Understanding the standard facility model helps drive decisions around investing or divesting in data centers. Furthermore, for some services, outsourcing and the emergence of cloud computing can sometimes lead to the decision to have the services hosted in external facilities.

All these standards must be documented so that the people that interact with them can easily access and apply them. This is aided by the creation of an architecture repository, a place where standards are stored and organized in a way that fosters collaboration among different stakeholders.

Key metrics for the standardization step

Capturing metrics in the standardize step provides an as-is snapshot and promotes effective governance of the transformation process. There are, of course, many metrics to choose from and it's impossible to create one set of metrics that are applicable to everyone. Here is a representative sample of some important metrics for standardizing:

- Number of standardized services targeted for the portfolio for the near-term
- Number of standardized services to date
- Current service operational cost expressed as a percent of the IT budget
- Targeted service cost

Implications and potential problem areas

The most significant "win" of the standardization step is the emergence of a value chain based structure and an enterprise architecture model and repository—even if it's just for a few services early on. These two elements are the foundational underpinnings of how the organization, its deliverables and its customers will interact moving forward.

Because standardization happens at the beginning of the transformation process, the major pitfall here is losing boardroom support before the benefits of transformation are realized. So it's important to begin with a single service, or small group of services, whose transformation will be seen as an early success and a harbinger of things to come.

With the standardization step complete, a working model has been created, and we can now begin the task of turning that model into reality.

Step 2: Instantiate

A McDonald's restaurant was recently remodeled in our neighborhood, a process that actually required tearing down the old one and building a new one on the same site. It was interesting watching it unfold because from the beginning it was clear that a model for the physical facility and a recipe for how to construct it were in motion. After what seemed like only several weeks, the restaurant was training their staff and testing out the facility. I went to the door and saw people, but they weren't open for business. Peeking in the glass, it looked just like one I had visited in another city. The interior was the same. It's quite clear that there is a pre-existing model for how to build or instantiate a McDonald's restaurant.

In this step we create the service we are transforming based on the pre-defined end state model and move it from the drawing board to actual implementation.

This is referred to as instantiating the service because it involves creating the revised or new models for the service that were defined in the standardize step, testing for function and quality, standing up the value chain based structure then moving to production.

The different perspectives of instantiation

Each value chain plays a role in the instantiate step. Let's look at some example activities:

- IT administrative management chain: setting up the financial chart of accounts for the service in the budgeting process, ensuring any changes to the chargeback model are made, coordinating with procurement to ensure the acquisition of new hardware, software or human resources to be applied to the service landscape

- BRM chain: building service test plans with the respective user communities, implementing the demand management process, prioritizing the migration order to the new service environment

- Service design chain: installation of any new technologies required for the service, standing up the service in the test environment and migration to production, ensuring design processes modeled in the standardize step are in place and working

- Service delivery chain: standing up and testing the various operational processes modeled for managing the service, engaging the new or modified production acceptance process, coordinating production acceptance with the BRM chain and the end user community.

For our web hosting service we would be installing all of the required hardware and software components of our BEA WebLogic offering. This might include racks of blade servers running Linux for the application and web tiers of the service and storage for our database tier. All of the network devices, circuits and gateways would be put in place and appropriately configured. We would then install and configure our WebLogic image, configure it and do some preliminary testing to validate the different aspects of our standard design. Once validated, we are ready to build out and implement the automated ecosystem we will use to manage the service—which is the next step of the recipe.

With the new service environment in place, it can be tested, refined and retested until it is ready to be moved *in total*, to the production IT environment. The operational model and processes, development standards and infrastructure architecture should be recalibrated as course corrections are made. Finance, security and related business policies should be revised as appropriate, as well as the operations management tooling and business controls related to the service based on the virtualized model.

In effect, IT has built the restaurant, but they aren't ready to take live orders from the customers. The instantiation stage is all about making sure players in various tribes are fully prepared to adopt new structure and behaviors as set forth in the standardize phase.

Key metrics for instantiation

Performance in this step is measured using several relevant metrics including:

- IT asset-to-support ratios, such as number of servers per system administrator

- Performance against targeted service cost

- Service quality targets as expressed through an SLA

Implications and potential problem areas

Once the instantiate step is complete, the service is ready to handle production workloads. The financial aspects of the service should be, by now, transparent both to IT and the lines of business. The IT operating model will be refined and adjusted to support the environment, and the impact of information security policies and regulatory controls on cross-geography or customer sharing will be minimized.

The biggest potential problem area in this step is falling back into old behaviors. For example, if virtualization techniques are employed for the service, this requires new ways of approaching design, delivery and management. While the standardize step is the place where these new behaviors and techniques are modeled, the instantiate step is where they are put into motion, tested and adjusted. Breaking old habits is difficult and often painful. Use the instantiate step as a catalyst for getting past the reluctance to change.

Step 3: Automate

In this step, we focus on applying hardware and software to the process and information models associated with the service(s) using the automation model from the standardize step. The goal in this step is to build out an integrated IT for IT ecosystem in a controlled manner that:

- Simplifies the systems architecture and diminishes tool complexity
- Reduces dependency on human intervention
- Lowers cost
- Improves predictability and speed
- Provides access to better information that can be used to facilitate better IT and business decisions

Most large IT organizations suffer from the same system and tool complexity, and data quality issues as their business units. So there are already a number of solutions in IT portfolio from which to choose to accomplish this goal. However, some gaps in the portfolio will need to be filled by making additional investments in software.

At a high level, the approach to automating IT should be similar to the one used to automate the value chains in virtually every other line of business. Using the designs from the standardize step, it becomes easier to apply standard systems development lifecycle methodologies and design techniques, such as model-driven automation or SOA, to IT automation.

As a reminder, automation needs exist within IT each value chain at three levels: task, process and service. With this in mind, let's spend a little time looking in detail at some examples of the automation opportunities and some priorities for this phase of the transformation.

Automating the IT administrative management chain

In this value chain, there are four areas of automation that warrant significant attention: finance, HR, portfolio management, and procurement.

For finance, automation should take place in all core processes: budgeting, accounts payable, service costing/pricing, and charge-backs. Most enterprises have a suite of administrative systems used by the lines of business that IT should leverage to support these wherever possible. However, many IT shops are ill equipped for activity-based costing which often accompanies a migration to service-centric IT. This makes it an example gap that might require new investment.

For HR, automation should target effective demand and supply matching. Improvements in the ability to manage and track the skills of the organization to support matching the right people (supply) with the prioritized projects (demand). Most enterprise HR systems focus on the dynamics associated with managing the workforce in general terms. In IT, however, management happens at a more granular level. Demand and supply matching is most commonly thought of in technology-related terms, such as: Is there enough server capacity?. However, having the appropriate level of human resources with the appropriate skills is a vital element in the service-centric demand and supply matching effort. In large organizations, tracking and managing skills usually isn't done well—which makes it another example gap that might require new investment.

For portfolio management, automation is most often associated with tracking the IT projects and their associated financial impact. However, since not all IT work is project related, we expand on this concept to become a systematic approach to managing major classes of items within the IT landscape. To get started, portfolio automation should focus on providing basic capabilities for:

- Project/program portfolio: includes managing the financial and human resources affixed to the projects activities surrounding the service focusing on the split between innovation and maintenance work
- Service portfolio: includes managing the service offerings and their attributes, as modeled in the standardize step
- Asset portfolio: including technology, applications and facilities
- Supplier portfolio: management of the vendor list and associated spending
- IT skill set portfolio: as referred to in the HR automation above, tracking this as a portfolio enables decision making around where to invest in training or even outsourcing

IT portfolio management enables the organization to quantify, manage and track IT investments and to enable fact-based decisions on future spending. Automating IT portfolio management produces a master data set that feeds the entire IT ecosystem.

For procurement, the areas of supplier and contracts management are ripe for automation. Automation in these areas can shorten delivery cycle time from suppliers via automated information exchange and improved contract information enables increased volume purchasing power, which optimizes cost.

For our web hosting service we would be automating the abilities to manage the financial aspects of the service including activities like the costing, budgeting and chargeback. We would also implement the portfolio management system to track the different attributes of this service like the service description (service portfolio/catalog), the associated assets (asset portfolio), any relevant projects (project portfolio) and skill set portfolio to track the resources with skills required for our service like BEA WebLogic, perhaps Linux and others. We would also make sure that the contract terms and conditions with BEA and other IT suppliers associated with the service were resident in a contract management system.

Automating the business relationship management chain

BRM chain automation includes enabling the SLA management, demand management process and requirements management, the latter two of which we will expand on a little further here.

For demand planning, automation, usually in the form of portfolio management software, helps establish forecasts for consumption of existing IT services and project requests for new services within the different lines of business. This demand is linked with pricing and financial targets to ensure that IT both delivers and derives maximum value from each service. The data associated with this vertical demand planning is assimilated into the IT administrative management value chain for horizontal rationalization, prioritization and investment planning.

For requirements management, software products are commercially available which automate the gathering and management of functional and process requirements from all stakeholder groups. Automation in this area is very important, so data associated with the service request can be visible to all requisite parties. This speeds the design process and ensures that deliverables match the original request.

While many automation opportunities in the BRM value chain are warranted, these two are highlighted as top priorities for all services as they pass through the transformation recipe. Additional automation should be applied as the transformation of subsequent services exposes additional gaps or opportunities.

For our web hosting service we would be automating the ability to capture requirements for new service requests and begin their lifecycle through an automated systems development lifecycle management system. We would also focus on ensuring the ability to capture upcoming demand for the service in the project portfolio implemented in the IT administrative chain automation. While not called out in detail above, this is also a good time to automate the service level management process to generate SLAs associated with the service.

Automating the service design chain

Service design chain automation priorities include: project management, service packaging, information management, and service testing.

For project management, automation helps control the progress of work, as well as monitor and measure performance against stated goals. Automation in this area also allows for time tracking, which measures the amount of effort divided between innovation and maintenance activities.

For service packaging, automation, usually in the form of portfolio management software and a service catalog, helps manage the service through its entire lifecycle. Automation here is targeted at turning the logical model from the standardize step into a physical reality. Implementation or enhancements to automation used to create physical databases, design, configure and deploy servers, storage and network are all part of this step. Furthermore, updating the service catalog resident in the service delivery chain with any new or changed service specifications is the responsibility of service design during this phase.

For information management (IM), the design chain owns the data warehouse in which information about the IT portfolios are held, including data about IT assets, people, services, and finance. In essence, the design chain hosts the data that will be accessed by the other supply chain stakeholders. The data design captured in the standardize step should serve as the blueprint for information management automation. IM automation supports IT business process management (BPM), integration, operational reporting and business intelligence solutions such as an IT data warehouse. Information produced can be used internally within IT and externally within the businesses to shape behaviors and guide fact-based investment decisions. The key is to create an authoritative source view of data within the IT ecosystem to ensure data quality, integrity and security.

Service testing automation can be implemented and augmented on a service-by-service basis as iterations of transformation recipe progress. Over 80 percent of all IT-generated business disruptions are caused by the introduction of changes to a given service. These changes are often a result of improper testing techniques. While there is no one-size-fits-all approach to service testing or testing automation, there are a finite number of patterns that exist in the testing process for each service line. These patterns should be captured and automated to the fullest extent as the services are transformed.

For our web hosting service we would be automating the abilities to package the service into an offering that can be made available to consumers and placing it into a service catalog system, asset management system and perhaps an image management system. We would also be installing automated testing systems to ensure consistency and accuracy of the service behavior as it moves through the production pipeline. Another important element of the automation in service design is the creation of a data warehouse environment to enable reporting on the performance, health and associated business value derived by our service.

Automating the service delivery chain

Service delivery chain automation priorities include:

- Order management, for enabling the consumer to buy the service and check fulfillment status
- Resource rationalization, for checking inventory of materials and people needed to produce the service
- Build/configure, for constructing the service using the necessary components
- Provisioning, for delivering the service to the buyer, and
- Management, including monitoring and recovery, applying the FCAPS model for fault, configuration, accounting, performance and security management.

For order management, solutions are commercially available that automate the service catalog, service order configuration and price quote along with controlling order approval routing. The service catalog will be the single pane of glass that the economic buyers use to engage with the IT function. Therefore, ease of use, accessibility and control are vitally important to selecting the service catalog solution. Quote and configure is a generic process used in most e-commerce marketplaces but is missing from the IT landscape. Using such automation in the service-centric IT organization helps improve the accuracy of the deliverables to the business. Finally, automated order approval routing is vital to ensure compliance with company policy around procurement of IT services.

For resource rationalization, automation combines data from IT asset management, skills management, configuration management database (CMDB) and perhaps parts inventory management applications to determine whether the appropriate resources are available to fulfill the service request. This type of solution is quite common in supply chain automation portfolios. However, in capability-centric IT organizations resource rationalization typically is performed manually, making this a likely gap that will need to be filled by new software.

For build and configure, automation typically targets tasks associated with the service environment. For example, automation can be used to increase network bandwidth, allocate additional space to databases, or instantiate an operating system configuration on a server.

For provisioning, automated change management solutions exist in all technology domains, simplifying the provisioning process and facilitating service-related changes anywhere in the IT environment. In addition, self-help automation solutions should be provisioned to decrease the number of help desk operators and to move knowledge closer to the service consumers.

For management, automating around the FCAPS model provides the most consistent and beneficial results to the service-centric IT organization. This is not a revolutionary concept—it's basic blocking and tackling for the IT teams in the service delivery chain. What's important at this stage is that consistent automated controls are put in place for the service being transformed, eliminating layers of redundancy and thus simplifying the management toolkit, reducing service maintenance cost, improving responsiveness of IT resources and improving IT data quality.

For our web hosting service we need to ensure the basic ability to monitor and manage the service components including the ability to discover, configure, change, identify and resolve problems with them. We would also automate the order management aspects of the service including configuring and quoting a price for the service, capturing an order, approval routing and if applicable, delivering the different requests for work to the appropriate resources for service construction. Because our service is highly standardized, we should be able to build and package, configure and provision the service through automated means, delivering technology capacity and software images for BEA WebLogic, operating systems, network device configurations, security policies and client configurations in a flexible and on-demand manner.

Key metrics for automation

Once in place, performance of the newly automated processes is measured using metrics such as:

- IT operation costs, including the cost of change or cost per help desk call
- Delivery cycle time
- Data quality (Six Sigma)
- Number of tools/systems per process
- Error rates in service delivery

Two emerging metrics in the automation domain are time-to-value, where IT measures the time it takes for development and subsequent deployment of projects; and time-to-productivity, which measures the amount of time and money required before IT professionals can be productive using the automated solutions.

Implications and potential problem areas

With so many opportunities for automation and so many different tribes, it's easy to see how redundancy and complexity can get out of control in the IT organization. This step helps build out the systems architecture for IT which serves as a governance tool for future iterations of the recipe. Seen in the context of the EA for IT, each automation element becomes part of a larger ecosystem (the ERP for IT), and can be accepted or rejected based on how well it fits into the structure and operating model of the IT business.

Automating functions within the value chains helps maximize cost effectiveness through the elimination of redundant licenses, maintenance agreements, hardware and software. Information in this automated ecosystem flows more freely, because data that traditionally stayed locked behind tools and applications now is part of an integrated information store that everyone, including the lines of business, can access. Getting the right information to the right people at the right time not only helps improve the effectiveness of IT, it also helps shape better investment decisions, operational excellence and behaviors in the business.

By completing the automate step in the transformation recipe, both the individual service and IT as a whole will take a big step away from the capability-based model and toward a service-centric model with automated processes.

The biggest landmine in this area is the emotional attachment that the tribe members have to their tools and solutions. So it's important to adhere to the authoritative model established in the standardize step when automating the service. This creates a fact-based blueprint for selection and implementation while eliminating redundancy and complexity.

Step 4: Consolidate

Now that the service structure and our enterprise architecture have begun to take shape as a result of standardizing, instantiating, and automating, we can shift existing and upcoming demand to this standardized, configure-to-order offering, which is where the payoffs start to become visible. The first three steps put the organization in a position to reap the benefits from the output of consolidation.

Consolidation improves the baseline operating cost for IT moving forward, and also drives resource utilization improvements for technology assets, facilities, and labor. For example:

- By consolidating into fewer, perhaps virtualized platforms, the utilization rates of the assets will move from the typical thirty percent metric to sixty percent as more services are transformed into a standards-based model. This helps avoid incremental investment in technology and simplify the IT architecture supporting the business.

- From a data center perspective, consolidation frees up floor space to help reduce power and cooling costs, and support growth from incremental business demand.

- From a labor perspective, consolidating into a highly automated, standards-based service environment drives down the number of employees required for support. In addition, it provides a better understanding of the economic envelope, which provides negotiating leverage when evaluating future investments, outsourcing or out-tasking decisions.

While consolidation is typically viewed as a technology event, we must not forget that it also has an impact on the structural aspects of the IT organization. This consolidation step is the act of reducing and simplifying the assets, processes, tools, and even people by moving them into the standardized service landscape. In our example of the hosting service, this means that we'll take any services or component capabilities that exist in the enterprise that match our BEA WebLogic pattern, and bring them into the fold of the transformed service environment. Thus, the capacity and the operational maturity for this standard service will expand while the existing build-to-order or non-standard hosting capabilities will be slated for retirement.

What follows are the basics of the consolidation step.

Develop a consolidation plan for each service

A robust consolidation plan should focus along vertical and horizontal dimensions to keep what should be a ballet from becoming a hockey match. The vertical aspect of the planning targets how and when the services will be consolidated in respect to each affected line of business—aligning what the business is capable of doing with the associated IT resource requirements in each value chain. This vertical planning not only helps the business itself get prepared, but helps them work with suppliers, partners or even customers to minimize risk and avoid confusion. Vertical consolidation plans are then rolled up, rationalized against staffing and funding, and prioritized horizontally to produce a comprehensive, interlocked plan of record that can be actuated for each service being transformed. Resources from each IT value chain play a role in developing an executable, service consolidation plan. To provide a flavor for the different planning perspectives, here are some examples:

- IT administrative management chain: Rationalizing the consolidation plans from the IT functions and the lines of business to prioritize work and forecast projected service demand to ensure that supply is readily available in the new service environment.

- BRM chain: Coordinating the development of the vertical plans within each respective line of business that subscribes to the service. Each plan should allow ample time for training and testing with the business not only of the service, but also the processes for ordering, reporting problems and requesting changes.

- Service design chain: Plans for performing architecture reviews to ensure standards compliance, such as for technology and security, for each migration package. This helps minimize the number of surprises as services flow through the consolidation pipeline. It also helps the service design engineers make any necessary course corrections to their pre-defined models from the standardize step.

- Service delivery chain: Plans developed for production acceptance of the migrating systems and services produced in the BRM and IT administrative management chains. These plans help the operational resources manage the new service environment using the new processes and tools, while at the same time maintaining the yet to be transformed service environments.

Execute the consolidation plan

This phase of the consolidation mobilizes the plans, moving them from the drawing board into action. In short, the residents are moving into their new homes in a new neighborhood complete with new rules and regulations. Move packages are constructed as resources begin working with the existing services and component capabilities. These move packages are staged or queued according to the service consolidation plan. How many or few consolidation move packages are in motion simultaneously is dependent on the capacity, digestibility and risk tolerance of the organization.

In our hosting example, a WebLogic based environment, perhaps resident in the Singapore data center, would be copied to the new service environment, perhaps hosted in Munich. Any adjustments required to conform to the standard model would be made and the testing and training portions of the consolidation plan would commence. Once complete, the signal for a final copy of the transactions from the existing environment would be taken and the failover to the new environment would take place. The service consumers would be using the service in the new environment with minimal or no disruption. The IT teams should be operating under the auspices of processes and tools put in place in the instantiate and automate phases of the transformation.

Once the consolidation of a specific service is complete, tested and verified by the consumers and IT, the next environment(s) in the queue is undertaken. This series of activities in the consolidation plans are performed until all of the pre-arranged services make their way to the new, standardized environment.

Each service consolidation has an affect on reducing and simplifying technology, people and process and ultimately delivers higher level benefits such as consolidation of facilities, an optimized labor pool and cost structure. Here are some examples of why this happens:

- From a technical perspective, there should be fewer moving parts in the IT landscape, simplifying the architecture and application portfolio. Fewer environments yield reduced data center expense, lower hardware depreciation and software maintenance costs.
- A simplified technology and application landscape requires fewer resources to manage. In addition, the standardization of the technology elements associated with the services drives a higher degree of utilization of the resources that remain. So in addition to a reduced labor line in the

budget, there is a higher degree of resource utilization and competency which leads to higher service quality and value delivered to the business.

- In addition, because of standardization, IT processes can be simplified which enables a streamlined automation capability resulting in fewer tools and more predictable delivery to the business. Fewer tools yields lower software maintenance and license costs and a higher degree of data quality associated with service performance.

Consolidation is hard work. Even for seemingly simple services, it requires collaborative efforts of IT and business resources managing a complex set of variables. The deliberate scope limits put in place in the transformation recipe, i.e. doing just one service to start, are there to help minimize the number of resources or teams dedicated to the effort and complexity to a relatively tolerable level.

With the consolidation of the services complete, we are now ready to proceed to the optimize step where we will retire the pre-existing services or capability components and make any adjustments to our transformation plans and models.

In our web hosting service example we would be concentrating on consolidating the existing BEA WebLogic-based environments around the world into the environment hosted in the Omaha data center. Each BRM team would develop move plans based the business calendars and priorities for their respective business functions. These move plans would be rationalized by a consolidation coordinator or project management office for the service and any potential collisions or conflicts would be rationalized and resolved, arriving at a comprehensive consolidation plan for the service. We would then capture the plan in the project portfolio management system and, once approved, scheduled, staffed and funded, begin execution. This execution would go through a series of migration, test and release motions until all existing environments, slated for consolidation, were complete.

Key metrics for consolidation

Example metrics to watch and control during the consolidate phase are primarily associated with service delivery and service design. These include:

- Speed of delivery: how long it takes to get a service from order to fulfillment

- Service quality: error rates in delivery and changes

- Help desk call volumes
- Cost per call
- Supplier costs: licenses, hardware depreciation, telecommunications costs

IT teams will also watch performance and utilization statistics and should ensure that measurements are taken before and after consolidation.

Implications and potential problem areas

The number one consolidation initiative in the IT arena today is data center consolidation. For large entities, this requires the coordination of a large number of resources and the orchestration of an extremely complex set of plans, not to mention an exorbitant amount of money. Because of the size, scope and complexity of data center consolidation efforts, there is a high risk of project atrophy, cost overruns and failure to realize the expected benefits.

Instead, using a service-centric approach to generate consolidation gives IT the ability to work in a deliberate and controlled manner, which yields results in a short time frame and allows for the organization to simplify data centers and technology while at the same time instituting new structure and behaviors that are the cornerstones of producing lasting benefit.

Two potential pitfalls in this step stem from simple human nature: first, trying to achieve one hundred percent success in the first pass instead of letting the effects of multiple iterations snowball over time toward completeness; and second, skipping over the first three steps and moving directly to consolidation to attain some form of instant gratification. However, without investing the appropriate amount of energy into the modeling of standard services, instantiating the environments and automating the activities, even the best consolidation plans can lead to lackluster results. The final pitfall stems from the common IT malady of letting the problem get too big. Keep it simple. Don't try to consolidate all services at once, remain focused on a few services at a time. To be successful, IT must stay focused on the mechanics underpinning the service-centric transformation recipe gaining confidence in the process as iterations occur.

Step 5: Optimize

By definition, optimization is about making something function at its best or most effective state. In the case of our recipe, the object of the optimize step is the newly transformed service and cleaning up the remnants of its journey. Using the enterprise architecture and the metrics from the four prior steps as a backdrop, optimization answers these basic questions: Is the transformed service delivering the expected benefits? Are there holes in the IT fabric left by consolidation that need to be repaired? Are there any service components that can be retired or repurposed? Do the structural models designed in the enterprise architecture reflect reality?

Answers to these questions generate activities that target recalibrating or adjusting areas in all IT value chains.

A value chain view of optimization

- IT administrative management: Making any necessary adjustments to the financial model for the service including budgeting, supplier re-negotiations, changes to charge-back or allocation models. Harvesting of the financial benefits is performed in the optimize step and any changes to the transformation plans are made. Any required changes to the IT organizational structure are made as well.
- BRM: Ensures that the transformed service is working as planned and delivering the expected business value. BRM validates IT's performance against SLA specifications and validates demand plans for the service(s) in future periods.
- Service design: Repairs any holes in the IT fabric created by consolidation. For example, if systems have been moved out of data centers, network redesign might be necessary or circuits rerouted. In addition, design chain resources measure the service in its final form against the original design, making any necessary adjustments to the enterprise architecture.
- Service delivery: Turns off or re-provisions assets rendered obsolete or re-usable by consolidation. Support processes associated with the service might be fine-tuned, and any necessary changes made to the associated automation based on actual service performance.

Optimization is not a one-time event. Instead, learnings from this phase help the organization to implement rigor and discipline that enable optimization to become a recurring or perpetual activity.

In the service-centric IT organization, perpetual optimization occurs in all of the value chains, tuning the technology, process, financial and even organization attributes of the service in accordance with changing business needs and demands throughout its entire lifecycle. Basic levels of this practice already exist in most IT organizations such as capacity management or performance tuning. However, most, if not all of the activities target the technical capabilities like servers, storage, network, etc. As more and more services flow through this step of the recipe, the perpetual optimization discipline and rigor become ingrained in the DNA of the organization—changing behaviors out of habit instead of edict.

With our BEA WebLogic service transformed, we can now go back and clean up the pre-existing environments. We would turn off any network circuits, servers and storage devices rendered surplus by our consolidation effort and do any necessary tuning of the IT infrastructure based on its new configuration. We would reclaim any assets deemed reusable for future iterations of the recipe and set them aside—moving the rest into the company's surplus asset removal process.

We would rationalize the contracts with BEA and other suppliers involved with the service environments to ensure compliance with terms or renegotiation or cancellation of existing contracts based on the new model. We would launch our staff recalibration effort, re-tooling those people we desired to retain and releasing those we did not. We would capture the financial savings directly associated with our service transformation such as reduced labor, utilities, software licenses, outsourced labor, and hardware maintenance. We would also capture any financial savings indirectly associated with our service transformation such as real estate fees (e.g. lease termination or data center property sell-off), space build out avoidance or perhaps any administrative maintenance expense savings. Finally, in our optimize step, we would do a post-mortem on the effort and expose any suggested course corrections that can be used for the next iteration of the recipe.

Key metrics for optimization

Because optimize is the final step in the recipe for a given service, its best to look at metrics here through a balanced scorecard view to establish new benchmarks for the organization. What follows are some examples of this balanced scorecard view of the metrics.

In terms of efficiency, the primary measurements in this step are associated with the performance of the servers, storage and network, to ensure that the optimized service environment complies with business needs, design goals, and SLAs.

In the finance area, a final snapshot of the service cost (or perhaps the cost per transaction) should also be captured in case any modifications were made in prior transformation steps. This service cost metric serves as the go-forward measure on which to base the economic value of the service to the business. Project costs associated with the service transformation should be captured in order to build accuracy into the financial forecast for the cost of passing subsequent services through the recipe.

From a customer perspective, service quality, time-to-response, mean-time-to-repair and cost-per-call metrics should be reviewed to assess the value associated with the transformation and to establish new benchmarks moving forward.

Implications and potential problem areas

Optimization can produce real economic benefits in the form of service contract eliminations and reductions with hardware, software, telecommunications and service suppliers. In addition, it produces cost avoidance opportunities through the re-purposing of existing assets and resources for other services being transformed, thus avoiding the need for additional investment.

The optimize step also provides a way for IT to evaluate any course corrections that should be made in the transformation plan. In addition, a foundation exists for the next iteration of the recipe. Service models, technology platforms, processes and automation solutions should serve as the baseline for the next set of services to be transformed.

The most common pitfall in the optimize step is to focus solely on technology, aka performance tuning. It is just as important to optimize the structure and operating model attributes of the service in this phase. Areas like engagement, design, support and finance should be recalibrated in relation to the transformed service. While this might make things a little more difficult in the short term, in the long run there will be convergence, and then simplification.

The Transformed Service

Taking a service through the transformation recipe should take no more than sixty to ninety days—perfect for demonstrating short-term results so that the executive suite and IT staff don't lose confidence, patience or passion for the transformation. So, especially early on, it's important to select services that can be transformed inside this time window. More complex services should be saved for later iterations so that most foundational elements are in place.

After completion of the first few recipe cycles, the shift towards a service-centric IT function should become visible to the business, the economic buyers of IT-delivered services. Service offerings should become easier to obtain and should be linked with measurable business value. Inside IT, there should be improved consistency in the technology that underlies each service, as well as a cleaner, simpler structure and operating model for how transformed services are managed through their lifecycle. The performance of transformed services against stated goals can be measured with real data, which makes SLAs more meaningful and drives improved decisions around service improvements. In short, the first few iterations of the recipe lay the foundation for the service taxonomy, the enterprise architecture for IT and the systems architecture that automates the function.

After a few trips through the recipe, transformation efforts should increase in scope and velocity, based on the ability to leverage the automation, taxonomy and structure from previous passes. This should enable IT to move more and more services through the transformation process simultaneously and at greater speed.

With each pass, there should be visible improvements to all key performance indicators, including speed, quality, cost and agility. Data centers will be consolidating, application portfolios will move towards simplification, IT supply will be able to flex with demand, labor costs will drop and speed of delivery will improve.

In addition, the IT organization itself will achieve a new level of maturity. Armed with the knowledge of the structure and operating model, IT professionals can better understand how their roles fit into the bigger picture enabling improved collaboration in design, delivery and ongoing management of services. This collaboration breaks through the tribal boundaries resulting in improved designs that are capable of delivering more innovation into the solutions that power the business.

Service-centric transformation delivers tactical and tactile progress towards the strategic end state that most "big rock" transformation initiatives fail to deliver. Because the effects of the transformation are visible to everyone, the boardroom discussions around IT shift from cost to value, from IT outcomes to the business outcomes, enabled by IT.

Summary

We've written this book to stimulate creative thinking and an active dialog around the changes needed within IT to help businesses embrace and compete in the service-centric era. Business and IT leaders we've talked with are at various stages of acceptance of and execution toward this concept. Some are using it as a means to mobilize and address long standing IT conditions such as cost control, complexity, alignment, governance and others. Others are continuing down the path of transitional programs such as consolidation or labor arbitrage, which they feel they must do to stay fiscally fit.

Regardless of which side of the acceptance ledger you fall, the data is clear that this trend will be disruptive to the normal course of business in IT sooner rather than later. Therefore, we urge readers to begin developing appropriate levels of readiness in order to avoid being caught off guard as the service-centric approach to IT takes shape.

The various constructs put forth in this book are not necessarily breakthrough on their own. In fact, the techniques and approaches discussed here are used daily by IT practitioners and leaders around the world. What's different is how IT can use these constructs to establish the right context and organizing principle for dealing with the speed of business change today.

A service-centric approach is a game changer, because it acknowledges that the service is the core deliverable, economic unit of exchange, the key organization element and a transformation catalyst that helps change all layers of the organization.

Using a service-centric approach is a game changer in the IT world, because it acknowledges that the service is the core deliverable, economic unit of exchange, the key organization element and a transformation catalyst that helps change all layers of the organization. The content of this book really can therefore be distilled down into four major themes:

Theme 1: The business landscape is changing to a multi-sourced, service-led model which will have a significant impact on how businesses are structured. The IT organization of today is not adaptable or scalable to support the changes needed to compete in this new market economy, and therefore IT must transform itself.

IT has become the central nervous system of the enterprise, woven into almost every aspect of the business fabric. IT departments are no longer just a supplier of applications, servers, data centers and help desks, but are at the very foundation of how the business strategy is executed—how goods and services are designed, manufactured and delivered to customers to drive profitable growth. And with this foundation in place, businesses around the world are moving toward new levels of competitive differentiation in a developing service-centric market economy using the internet as a platform for their business services, accessible from a multitude of different devices.

Whether you call it cloud computing, service-oriented architecture or utility computing, the fact is that a new era of information technology is emerging, bringing with it new opportunities for business breakthroughs, while at the same time introducing a new set of challenges for CIOs and IT professionals.

This service-centric model will put more pressure than ever on IT to control both *what* is delivered to the business and *how* it's delivered. Given that services can be obtained from anywhere, at anytime, by anyone, the task of managing the business technology ecosystem will be vastly different than today's internally sourced model. Sourcing of business services, governance, portfolio management, securing company and customer data and ensuring business operations stability on a daily basis will become even more complex.

It's clear that the methods for running the IT function in this new era will need to be different than today's version. For example, in addition to focusing on internally sourced infrastructure and applications, IT will have to contend with how to manage external or multi-sourced services from a wide array of providers.

The legacy IT operating models, steeped in its tribe-like structure, will not flex and scale to meet the needs of the enterprise of the future. Therefore, most business and IT leaders are looking at how to transform the IT function to accommodate and embrace this new set of challenges.

This is not to say that the pre-existing issues are no longer valid. CIOs continue to battle with overburdened staff, shrinking budgets and an increasing demand portfolio for new capabilities in the face of this transformation demand. The challenge is how to do both—execute the transformation while at the same time delivering to existing business needs.

Theme 2: IT Transformation to the service-centric model is more than just hardware, software or data center changes. The initiative should become part of the business strategy and should be framed in a manner that targets three key areas simultaneously:

- Establishing services as the economic unit of measure between the business and IT: Services are the products or deliverables of IT and they may be sourced internally or externally. The key to success is modeling them in a manner that delivers a clear picture of the cost envelope and to call out how, when and where they deliver business value.
- Creating a structure for IT based on the services: Organizing IT around value chains set in the context of the services being offered helps break through much of the silo mentality that exists today. That's not to say that silos are eliminated, instead it's about forcing collaboration to support the service through its lifecycle. Using an enterprise architecture framing for governing and communicating IT structure is a good way to help resources in the business and IT understand how things work and to govern change.
- Leverage an integrated software fabric to automate the processes and information flows in each value chain. Just like any line of business depends on its systems and information to conduct business, so does IT. The fact is that there is an overabundance of tools that have propagated throughout the IT tribes and these need to be consolidated into an ERP like ecosystem for the function. Using the EA for IT as a guide and best practice frameworks like ITIL as the process and service lifecycle framing, the systems architecture for IT can be constructed in a deliberate manner instead of an opportunistic one.

What we've tried to provide with this framework is guidance to aid in transforming IT from a capability-centric function to a service-centric line of business—moving from an expensive, build-to-order model to a more agile and efficient configure-to-order model. Transformation is a wholesale, comprehensive change from the current means of running the IT business, and as such, it is hard work. Because IT is now part of the business fabric, it will require involvement from all corners of the enterprise. In order to ensure that the efforts don't atrophy or deviate from the course, we are advocating our pillars be used to guide the efforts. The four pillars in the plan are designed and positioned to cover the major touch points of successful transformation—its services, its structure, its automation and a recipe to guide progress.

The transformation framework we've laid out in this book is a starting point and we readily acknowledge that as successful transformations unfold, there might be enhancements and course corrections added. One point worth reiterating here is that the transformation plan is set in the context of several strategic criteria:

- The business strategy: what will the business look like on the other side of transformation? The IT strategy must be interwoven into the business strategy, not viewed as the CIO's project. This can be modeled using the company's enterprise architecture and is typically measured and expressed using standard metrics such as profit, growth, market segments, etc. In addition, however, we'd like to advocate examination and inclusion of IT-related metrics such as: number of processes per function and ratio of applications to processes so that the connection between IT transformation and business changes are maintained.

- Financial targets: what will the IT spending profile look like on the other side of transformation? How much will be spent on maintenance, how much on innovation? How much will be fixed cost, how much will be variable? Will the overarching IT financial KPI be expressed as a percent of revenue or will it be measured some other way? How will business benefits from IT projects and services be captured going forward? How will demand for IT services be prioritized?

- Headcount targets: what will the employee portfolio look like on the other side of transformation? Assuming that headcount can be reduced through standardization and consolidation, what are those targets? How much will come out year over year? What will the profile look like from a skills perspective? Where will the employees be located? What will be the permanent to contractor headcount mix be?

- Application targets: assuming that there is already an overabundance of business applications going into transformation, what will the portfolio look like on the other side? How many applications will we end up with? Which ones are strategic and which ones will be eliminated?

- Project portfolio targets: assuming again, there is already more work in the system than can ever get done, there needs to be a deliberate plan to get the funnel into a controllable state. What is the right number of projects given the targets that have been established above including the transformation work? How will work be prioritized—will it be vertically within each line of business, or horizontally across the company, or both?

Establishing these targets helps set the stage for positioning the transformation efforts in the context of the entire business strategy. Failure to do so will likely lead to atrophy of the transformation initiative and thus lead the IT organization right back to the starting gate.

Theme 3: The recipe actuates and guides progress for the transformation journey.

Transformation takes time. Most IT leaders who have undertaken this kind of initiative agree that it can take three years or more to complete. This is not a single-threaded project that will run its course in a month. Unfortunately, most businesses want near term benefits. The good news is that by using an iterative recipe, benefits can be realized in a short time horizon. And, as iterations of the recipe occur there is the creation of a transformation snowball that delivers compounding benefits throughout the lifetime of the journey.

This recipe should be nimble, yet comprehensive enough to iteratively build out the ERP for IT and the EA for IT. As with any effort, there are multiple approaches. The right approach for your organization will depend on the positioning of IT in the business strategy and the risk tolerance of the company. For example, some CIOs will be given the funding and latitude to drive a green field approach, perhaps building out new data centers and transforming into them, while others will be forced to work with constraints such as economic or staffing levels. Regardless of which one depicts your position, deciding on what to drive through the recipe is critical.

We have pointed out in this book that using a service-led transformation recipe is cleanest because it helps drive the technology, sourcing and organization shifts in a more controllable or measured manner. And, it helps contain what we refer to as the "IT attention deficit disorder" syndrome by keeping the focus area small and manageable. It also helps deliver short-term value and progress toward a strategic end state. As we covered in the prior sections, there are five basic steps to our recipe:

- Step one of the recipe focuses on standardization. By now it should be clear that standardization is more than a technology related topic. Standardization in a service-led IT transformation focuses on the technology, but it also focuses on standardizing the organization structure, cost models, engagement models, policies, processes, sourcing and automation architecture around a given service. Modeling a service is imperative to driving out redundancy, discerning business value and making fact based sourcing decisions. It also creates a governance anchor point on which supply and demand can be measured and value can be derived.

- Step two of the recipe focuses on instantiating the model. This is more than standing up a shared server farm. It's also about standing up the processes, organization structure and governance required to make the service real. IT practitioners are tactile beings, if they can't see it or manage it, it doesn't exist. Therefore, instantiating the service helps mobilize the workforce around something that is "real". From this step, the rest of the pieces of the puzzle can be put into place.

- Step three of the recipe focuses on automation. With an understanding of the service in hand, and a demonstrable proof point in place, there are enough facts on which to base requirements for a software-driven automation fabric that enables management of the service across the different value chains. Most IT functions will have tooling already in place that can be easily adopted and aligned with the service. However, integration across the value chain networks is key to managing the service through its entire lifecycle. We want to restate here that being perfect in the early stages of the recipe is not necessary. Iterations of the recipe should deliver a good enough level of automation so that subsequent passes extend towards completion.

- Step four of the recipe focuses on consolidation. With a viable service available the organization can begin migrating into the new service environment. This is where the first financial benefits become visible. By migrating into a consolidated environment there will be fewer things to manage, less floor space required and fewer resources involved in maintenance. This leads to recalibration opportunities in terms of sourcing options, staffing and funding levels—most of which will ultimately fall through to the bottom line financials of the company.

- Step five of the recipe is around optimization. This is another place to harvest financial benefits. Activities here ensure that clean up occurs to plug holes in the technology landscape and ensuring the organization is trained, aligned and engaged in working in a service led model. It is the step where assets are harvested for re-purposing, buildings are shutdown and staffing is reassigned. It is also important to use this step as a place for planning the next iteration(s) and applying lessons learned.

This recipe is not set in stone—however, there is a deliberate ordering to these steps so that IT and the business can deal with management of change issues brought on by transformation and IT can deliver results while at the same time driving significant change. While there is not a one size fits all recipe, these steps are common practice inside most IT transition or transformation efforts today. However, most of these steps are often occurring in a disorganized or perhaps accidental manner. Applying the recipe using the service as the baseline, the iterations should net out the benefits in a controllable, predictable manner.

Theme 4: A new IT emerges, a service-centric line of business equipped and able to deliver valued business outcomes.

A service-centric IT function has three major characteristics that combine to deliver value:

- It is predictable, easily engaged and transparent from a cost and value perspective. It has a well defined governance model, strong portfolio management regimen and sources services from multiple suppliers. Because of standardization, services can be rendered in a configure-to-order or on-demand model, reducing the number of build-to-order solutions that consume time, effort and money within IT.

- It is comprised of people that understand their line of business, how its structured and how things are supposed to work. As the action in the business heats up, these resources can be assigned to meet the changing needs of the business. This is real agility. The ability to accurately provide this level of demand and supply matching is only available to those companies that control both what is managed (the service) and how its managed (the ecosystem).

- It is aligned with the business. Real business alignment doesn't happen at the project or process level. It happens by ensuring that the operating model and structure of IT is constructed in the context of the structure and operating model of the business.

With the market economies moving to the service-centric model, IT must re-position itself to continue to drive competitive business differentiation into the enterprise. Without this transformation external sources of services will become more prevalent in the business, thus creating even more shadow IT, fragmentation and complexity. As with all pendulum swings or eras of computing, this one will have its extremes and will settle somewhere in the middle. However, we're hopeful that this book serves as a guide to help form the service-centric IT organization in its early stages.